SHG Approach in Improving Livelihood and Empowerment of Rural Poor

NIPA® GENX ELECTRONIC RESOURCES & SOLUTIONS P. LTD.
New Delhi-110 034

About the Authors

Banshi Lal Sahu, born on Twenty ninth day of July Nineteen Hundred and Ninety One, is presently working as Block Technology Manager, Agricultural Technology Management Agency (ATMA), Kanker, Chhattisgarh. He has done his B.Sc. (Ag.) Hons. from Indira Gandhi Krishi Viswavidyalaya (IGKV), Raipur in 2013 and M.Sc. (Ag.) in Agricultural Extension from Institute of Agriculture, Visva-Bharati, Sriniketan, West Bengal in 2015. He served as Agricultural Officer at *Raigarh Sahayog Samiti*, a Non-Government Organisation.

Dr. Souvik Ghosh, born on First day of November Nineteen Hundred and Seventy Two, is presently working as Professor (Agricultural Extension) at Deptt. of Agril. Extension, Agril. Economics and Agril. Statistics, *Palli Siksha Bhavana* (Institute of Agriculture), Visva-Bharati University. Earlier he worked as Scientist (Agricultural Extension) and Senior Scientist (Agricultural Extension) at ICAR-Indian Institute of Water Management, Bhubaneswar during July 2000 – June 2009 and July 2009 – June 2013, respectively. He graduated as Gold Medalist in B.Sc. (Agril.) Hons. from Visva-Bharati University, Santiniketan in the year 1995 and passed M.Sc. (Dairying) in Dairy Extension Education with Director's Gold Medal at National Dairy Research Institute, Karnal in 1997. After completion of Ph.D. (Agricultural Extension) from Indian Agricultural Research Institute, New Delhi with distinction of Significant Post-Graduate Research in 2000, Dr. Ghosh joined Agricultural Research Service (ARS) of Indian Council of Agricultural Research (ICAR) on 10th July 2000. His outstanding work done in the field of Agricultural Extension Research and Education has brought many laurels to him including Lal Bahadur Shastri Outstanding Young Scientist Award of ICAR, ICAR Outstanding Team Research Award, Fellow of Indian Society of Extension Education (ISEE), New Delhi, ISEE Young Scientist Award, Young Scientist Award of Society of Extension Education (SEE), Proficiency Award of Directorate of Water Management (DWM), Institution Award of Institution of Engineers (India)- Orissa State Centre, etc. A total of 225 publications of Dr. Ghosh including 92 papers in national and international journals and three books reflect the salient works accomplished by him. For his book on "वेशवीकरण और भारतीय खाद्य सुरक्षा" - Globalisation and Indian Food Security (ISBN-978-81-910947-0-1), *Indira Gandhi Rajbhasha Puraskar* (First Prize) of Ministry of Home Affairs, Govt. of India conferred by His Excellency President of India, Shri Pranab Mukherjee at Vigyan Bhavan, New Delhi on 14th September 2012.

SHG Approach in Improving Livelihood and Empowerment of Rural Poor

Banshi Lal Sahu
and
Souvik Ghosh

NIPA® GENX ELECTRONIC RESOURCES & SOLUTIONS P. LTD.
New Delhi-110 034

NIPA® GENX ELECTRONIC RESOURCES & SOLUTIONS P. LTD.

101,103, Vikas Surya Plaza, CU Block
L.S.C. Market, Pitam Pura, New Delhi-110 034
Ph : +91 11 27341616, 27341717, 27341718
E-mail: newindiapublishingagency@gmail.com
web: www.nipabooks.com

For customer assistance, please contact
Phone: + 91-11-27 34 17 17
Fax: + 91-11- 27 34 16 16
E-Mail: feedbacks@nipabooks.com

ISBN: 978-81-19254-41-5

Composed and Designed by NIPA®.

Preface

Rural poor including the women in India are influenced by multiple socio-economic and cultural factors. Emancipation of women is a pre-requisite for nation's economic development and social upliftment. The role of women and the need to empower them are central to human development programmes, including poverty alleviation. Over the last few years, 'People's participation' and 'Empowerment' has become the buzz words in rural development and local planning. In these contexts, self help group has emerged as the most successful strategy, in the process of participatory development and empowerment of rural poor including women.

The SHG is a viable organized setup to disburse micro-credit to the rural women for the purpose of making them enterprising and encouraging to enter into entrepreneurial activities. The formation of SHG is not ultimately a micro-credit project but also an empowerment process. The empowerment through SHGs would give benefits not only to the individual but also for the family and community as a whole through collective action for development. These SHGs have a common perception of need and impulse towards collective action. In this context, this book provides an account on the concept of SHG approach, group dynamics, livelihood and empowerment. A glimpse of past research on SHG approach in this book would help the future researchers, scholars, policy makers and practitioners. SHG approach in eastern Indian states with special emphasis on Chhattisgarh has been narrated with the detail of research methodologies to study SHG approach and their applications in analyses of twelve SHGs formed under four different programmes *viz.* NABARD's SHG-Bank Linkage programme (SBLP), National Rural Livelihood Mission (NRLM), Integrated Watershed Management Programme (IWMP), Agricultural Technology Management Agency (ATMA). This book also reflects on the integrated methodologies to understand overall impact of SHG approach on empowerment and livelihood of rural poor including the women. The concept of group dynamics and its assessment is another important dimension that is discussed in this book.

The dissertation work on the effectiveness of SHGs in improving livelihood and empowerment of rural poor in Chhattisgarh carried out during 2014-15 at Department of Agricultural Extension, Agricultural Economics and Agricultural Statistics, Institute of Agriculture, Visva-Bharati, Sriniketan by the first author in partial fulfilment of the requirements for the degree of Master of Science (Agriculture) in Agricultural Extension under the supervision of second author has been presented in this book to give an understanding on integrated approach to study the SHGs. Authors gratefully acknowledge the Visva-Bharati University for this study. The works/ information of other researchers/ experts/ sources referred/ included in this book have also been thankfully acknowledged.

This book would certainly assume great significance in creating database for realistic planning and implementation of future SHG movement and would help in adding to existing store-house of knowledge concerning SHG approach and related issues. It would also guide readers in deriving insight in understanding many aspects relevant particularly to SHG approach adopted under different schemes implemented by various government and non-government organizations.

Authors

Contents

Chapter 1

Concept of SHG Approach, Group Dynamics, Livelihood and Empowerment

The concept of Self Help Group (SHG) has its roots in rural areas and it has been mooted along the rural poor to improve their living conditions. Though it is applicable to both men and women in our country, but it has been more successful among women and they can start economic activities for improving their livelihoods through SHG movement. Self Help Group (SHG) is a small voluntary association. It is informal and homogenous group of not more than twenty members. SHGs consist of maximum 20 members because any group having more than 20 members has to be registered under Indian legal system. That is why, it is recommended to be informal to keep them away from bureaucracy, corruption, unnecessary administrative expenditure and profit motive. In fact, it is a home grown model for poverty reduction which simultaneously works to empower and shape the lives of its members in a better way. Groups are expected to be homogenous so that the members do not have conflicting interest and all the members can participate freely without any fear. SHG movement has triggered off a silent revolution in the rural credit delivery system in India. SHGs have proved as an effective medium for delivering credit to rural poor for their socio-economic empowerment. In the development paradigm, SHG and micro-finance has evolved as a need-based programme for empowerment and alleviation of poverty to the so far neglected target groups (women, poor, deprived, etc.).

In India, this scheme is implemented with the help of National Bank for Agriculture and Rural Development (NABARD) as a main nodal agency in rural development. It is self employment generation scheme for especially rural poor, who don't have their own assets. The NABARD introduced a pilot project commonly known as SHG linkage project in 1992. With an addition to this, in 1993, the Reserve Bank of India (RBI) allowed SHGs to open saving accounts in banks and avail the banking services and it was the major boost to the movement. With a small beginning in 1992 as a pilot project, the active participation of Government, Banks, Development Agencies and Non-Government Organisations (NGOs) has made the SHG movement in India as the world's largest microfinance programme.

SHGs have the power to create a socio-economic revolution in the rural areas of our country. SHGs have not only produced tangible assets and improved living conditions of the members but also, helped in changing much of their social outlook and activities. SHGs have served the cause of women empowerment, social solidarity and socio-economic betterment of the poor (Ramachandran and Balakrishnan, 2008).

When the bottom of the pyramid i.e. four billion people are converted into micro producers, opportunity for global growth becomes obvious. The real effectiveness and success depends on alleviating poverty by converting the poor into producers which will increase the income of the rural families (Rajendran, 2012).

SHGs have been identified as a way to alleviate poverty and women empowerment. And women empowerment aims at realizing their identities, power and potentiality in all spheres of lives. But the real empowerment is possible only when a woman has increased access to economic resources, more confidence and self motivation, more strength, more recognition and say in the family matters and more involvement through participation. Although it is a gradual and consistent process, but women should build their mindset for taking additional effort willingly for their overall development. SHGs have the potential to have an impact on women empowerment (Narang, 2012).

To reduce poverty by enabling the poor household to access gainful self employment and skilled wage employment opportunities, resulting in appreciable improvement in their livelihood on a sustainable basis, through building strong grass-root institutions of the poor (SHGs), is now the main motive of the most employment schemes. Thus, SHGs have been showing the way ahead to alleviate the poverty of India along with women empowerment.

1.1 Genesis and Growth of SHG Approach in India

Micro finance movement involving SHGs initiated during 1976 at Bangladesh by Mohammed Yunus. In the eighties, it was a serious attempt by the Government of India to promote an apex bank to take care of the financial needs of the poor, informal sector and rural areas. NABARD took steps during that period and initiated a search for alternative methods to fulfil the financial needs of the rural poor and informal sector. Real effort was taken after 1991-92 from the linkage of SHGs with the banks. In other words, the SHG in India has come a long way, since its inception in 1992. The spread of SHGs in India has been phenomenal.

It has made dramatic progress and 90 per cent of these groups are only women groups (NABARD, 2005). The NABARD declared that more than 400 women join the SHG movement every hour and an NGO joins the Micro-Finance Programme every day. The small beginning of linking only 500 SHGs to banks in 1992, had grown to over 0.5 million SHGs by March 2002 and further to 8 million SHGs by March 2012. Together the 8 million SHGs maintain a balance of over Rs. 6550 crore in the savings bank accounts with the banks, while they are estimated to have harnessed savings of over Rs. 22000 crore of which nearly 70% (over Rs.15000 crore) goes for internal lendings (NABARD, 2012).

The spread of the SHGs is highly concentrated in the southern part of the country with very few in the north and the east. Over half a million SHGs have been linked to banks over the years but in a handful of States, mostly in South India, account for more than 60%. Andhra Pradesh has over 42% of the total SHGs. Since the advent of SHG in India, its growth rate has been low in the states of Rajasthan, Bihar, Utter Pradesh, Madhya Pradesh, Orissa and union territory of Andaman Nicobar Islands where the status of women is still very backward and pathetic. The formation of SHGs have benefited its members in numerous ways; not only have the assets, incomes and employment opportunities for the women but also enhance the equality of status of women as participants, decision-makers and beneficiaries in the democratic, economic, social and cultural spheres of life (Jain, 2003).

Majority of the SHGs are prevalent in South Indian States. However, the purpose of SHG approach is more important in the Eastern Indian States, which are poverty stricken. Among the Eastern Indian States, Chhattisgarh is having relatively more percentage of rural households with at least one person belonging to a farmer-based organiz-ation and/or self-help group (Fig. 1.1). Therefore, an investi-gation, carried out to under-stand the effective-ness of SHGs in improving livelihood and empowerment of rural poor in Chhattisgarh during 2014-15, is presented in detail in subsequent Chapters.

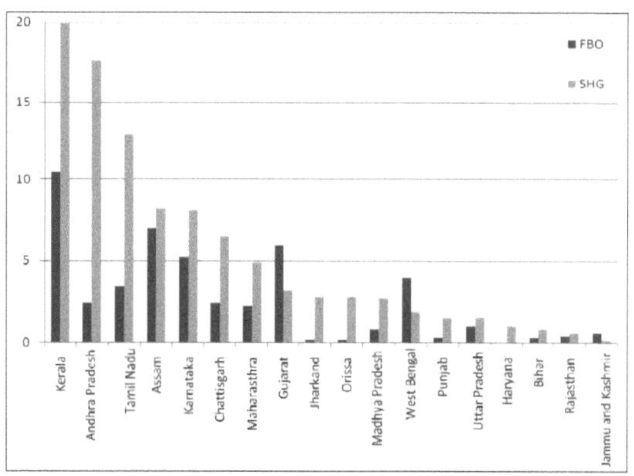

Fig. 1.1: Percentage of farmer households with at least one person belonging to a farmer-based organization and/or self-help group

(*Source*: Birner and Anderson 2007. Note: FBO: farmer-based organization, SHG: self-help group).

1.2 SHG: Developmental Paradigm

Villages are faced with problems related to poverty, illiteracy, lack of skills, health care, etc. These are problems that cannot be tackled individually but can be better solved through group efforts. Today these groups known as Self help groups have become the vehicle of change for the poor and marginalized.

Self-help group is a method of organising the poor people and the marginalized to come together to solve their individual problem. The SHG method is used by the government, NGOs and others worldwide. The poor collect their savings and save it in banks. In return they receive easy access to loans with a small rate of interest to start their micro unit enterprise. Thousands of the poor and the marginalized population in India are building their lives, their families and their society through Self help groups. The 9th five year plan of the government of India had given due recognition on the importance and the relevance of the Self-help group method to implement developmental schemes at the grassroots level.

- SHG is a development group for the poor and marginalized
- It is recognized by the government and does not require any formal registration
- The purpose of the SHG is to build the functional capacity of the poor and the marginalized in the field of employment and income generating activities
- People are responsible for their own future by organizing themselves into SHGs

The strong belief by the individual to bring about change through collective efforts:

- Effort is built on mutual trust and mutual support
- Every individual is equal and responsible
- Every individual is committed to the cause of the group
- Decision is based on the principles of consensus
- The belief and commitment by an individual that through the group their standard of living will improve
- Savings is the foundation on which to build the group for collective action.

A variety of group-based approaches that rely on social collateral and its many enabling and cost-reducing effects are a feature of modern microfinance (MF). It is possible to distinguish between:

- Groups that are primarily geared to deliver financial services provided by microfinance institutions (MFIs) to individual borrowers; and
- Groups that manage and lend their accumulated savings and externally leveraged funds to their members.

While the term 'Self-help group' or SHG can be used to describe a wide range of financial and non-financial associations, in India it has come to refer to a form of Accumulating Saving and Credit Association (ASCA) promoted by government agencies, NGOs or banks. Thus, SHGs fall within the latter category of groups described above.

Types of self-help groups engaged in financial intermediation in India distinguished by their origin and source of funds:

(i) Pre-existing groups:

- Rotating Savings and Credit Associations (ROSCAs), such as *nidhis*, rotating their own savings with no external resources
- Pre-existing ROSCAs that have been identified by banks and are accessing bank loans.

(ii) Promoted by NGOs:

- Thrift groups receiving no external funds (including those formed under component programmes of sector development projects)
- Receiving only revolving fund grant from NGO/donor
- Started with donor/government grants and subsequently linked to banks/ MFIs
- Receiving loans from NGO-MFI.

(iii) Promoted by Banks/Non-Banking Finance Companies (NBFCs):

- Promoted by bank staff and agents – receive loan from bank
- Promoted by agents – receive loan from bank/NBFC.

(iv) Promoted by District Rural Development Agency (DRDA)/Government/ Local Government agency:

- Only receive revolving fund grant from government agency
- Access loans from banks/MFIs.

(v) Promoted by SHG federations.

- Directly linked to banks
- Receive loans through SHG federations.

A distinction can be made between different types of SHGs according to their origin and sources of funds. Several SHGs have been carved out of larger groups, formed under pre-existing NGO programmes for thrift and credit or more broad-based activities. Some have been promoted by NGOs within the parameters of the bank linkage scheme but as part of an integrated development programme. Others have been promoted by banks and the district rural development agencies (DRDAs). Still others have been formed as a component of various physical and social infrastructure projects. Some of the characteristic features of SHGs are given below:

- An SHG is generally an economically homogeneous group formed through a process of self-selection based upon the affinity of its members.
- Most SHGs are women's groups with membership ranging between 10 and 20.
- SHGs have well-defined rules and by-laws, hold regular meetings and maintain records and savings and credit discipline.
- SHGs are self-managed institutions characterized by participatory and collective decision making.

NGO-promoted SHGs were often nested in sanghas or village development groups undertaking integrated development activities. As they have developed, SHGs or sanghas have been grouped into larger clusters and multi-village federations for financial and nonfinancial activities.

The basic concept remaining precisely the same, some additional features and working resilience have been incorporated in the SGSY-SHGs for making the concept still more user-friendly and for achieving the avowed objectives of SGSY in their true spirit. The notable features are given below:

	Features of SHGs
A. Strategy	Conceived as a holistic programme of self-employment. It covers all the aspects of self employment of the rural poor, viz. organizing them in SHGs, their capacity building selection of key activities, planning of activity clusters, infrastructure build-up, technology and marketing support.
B. SHG Formation	Specifically SHG members from BPL families with some exception for new marginal APL families if acceptable to the BPL members of the group.Group size of 10-20 persons, with the exception of deserts, hills and disabled persons where the number of members may vary from 5 to 20.Special focus on the formation of exclusive women Self-Help Groups. 50% of the groups formed in each block should be exclusively for women.SHGs are normally formed by NGOs, CBOs, Animators, Network of Community-based Coordinators, or team of dedicated functionaries of the government.
C. Income-Generating Activities (Micro-enterprise selection)	SGSY Committee identifies about 8-10 farm and non-farm key activities per block for the individual/SHG Swarozgaris of the block to choose some of them as the sustainable income- generating activity for themselves.Primarily, single income-generating activity by the group is given preference under group loan. Group, however, may go for multiple activities also under group loaning. Thus, IGAs are taken by the SHG members as a group activity.The focus is on the development of activity cluster to facilitate forward and backward linkages to IGAs.
D. Promotional Support	Revolving Fund Assistance (RFA) is provided to groups equal to their corpus within the prescribed limit.group corpus within the prescribed limit.
(i) Financial Support	Back-ended subsidy to the extent of 30% to 50% of the project cost is provided to individual beneficiary, and 50% of the project cost for group level activity is provided within the prescribed limits.
(ii) Group Formation and Nurturing Support	Financial Assistance is provided to NGOs/CBOs/SHPI, etc., for formation and development of SHGs, as mentioned below.Rs10,000 per SHG is paid for the formation and development of SHGs in four instalments. 1st – 20% at the beginning of the group formation. 2nd – 30% when group qualifies for Revolving Fund. 3rd – 40% when group takes up economic activity. 4th —10% after the start of economic activity and on adherence of group to repayment of bank loan.
(iii) Capacity Building Support	Fund support is made available to organize training of beneficiaries in group processes and skill development.
(iv) Infrastructure Building Support	There is planned focus on infrastructure build- up, technology and marketing support to make self-employment activity economically sustainable.

Source: VOICE (Voluntary Operation in Community & Environment) (2008). A Report on the Success and Failure of SHG's in India – Impediments and Paradigm of Success. Submitted to Planning Commission, GOI, New Delhi

1.3 Programmes following SHG Approach

The SHG movement was mooted by NABARD in early 1990s through SHG-Bank Linkage programme. The success of the SHG-Bank Linkage Programme (SBLP) has motivated the Government to borrow its design features and incorporate them in their poverty alleviation programme. Subsequently, many government organizations and non-government organizations have undertaken the SHG approach under various schemes like National Rural Livelihood Mission (NRLM), Integrated Watershed Management Programme (IWMP), Agricultural Technology Management Agency (ATMA), etc. This is certainly welcome, however has raised concerns regarding the process through which groups are formed under different programmes. There is a need to bring better synergy between these programmes. Therefore, effectiveness and functioning of SHGs formed under different such programmes will provide further impetus to the future planning on empowerment of rural poor.

1.3.1 NABARD's SHG-Bank Linkage Programme (SBLP)

The SHG – Bank Linkage Programme was started as an Action Research Project in 1989 which was the offshoot of a NABARD initiative during 1987 through sanctioning Rs. 10 lakh to MYRADA as seed money assistance for experimenting Credit Management Groups. In the same year the Ministry of Rural Development provided PRADAN with support to establish self-help groups in Rajasthan. 7.03 The experiences of these early efforts led to the approval of a pilot project by NABARD in 1992. The pilot project was designed as a partnership model between three agencies, viz., the SHGs, banks and NGOs. This was reviewed by a working group in 1995 that led to the evolution of a streamlined set of RBI approved guidelines to banks to enable SHGs to open bank accounts, based on a simple "inter se" agreement. This was coupled with a commitment by NABARD to provide refinance and promotional support to banks for the SHG-Bank Linkage Programme. 7.04 Initially there was a slow progress in the programme up to 1999 as only 32,995 groups were credit linked during the period 1992 to 1999. Since then the programme has been growing rapidly. SHGs are operating as thrift and credit groups. More than 90% of the members of SHGs are women and most of them are poor and assetless. The SHG movement has been instrumental in mainstreaming women by-passed by the banking system.

1.3.2 Swarnajayanti Gram Swarojgar Yojana (SGSY)

Swarnajayanti Gram Swarojgar Yojana (SGSY) has been launched with the objective of bringing every assisted family above the poverty line within three years, through the provision of micro enterprise. SGSY aims at establishing a large number of micro-enterprises in the rural areas, building upon the potential of the rural poor. It is rooted in the belief that the rural poor in India have competencies and, given the right support can be successful producers of valuable goods/services. SGSY is conceived as a holistic programme of micro-enterprises covering all aspects of self-employment viz. organisation of the rural poor into self-help groups (SHGs) and their capacity building, planning of activity clusters, infrastructure build up, technology, credit and marketing. The SGSY is designed to incorporate the following three important approaches:

Cluster Approach: The emphasis under SGSY is on the cluster approach. For this, 4-5 key activities are identified for each block based on the resources, occupational skills of the people and availability of markets. Selection of the key activities is with the approval of the Panchayat Samitis at the block level and the DRDA/ Zilla Parishad (ZP) at the district level. The major share of SGSY assistance is in activity clusters.

Project Approach: SGSY adopts a project approach for each key activity. Project reports are prepared in respect of identifies key activities. The banks and other financial institutions are closely associated and involved in preparing these project reports, so as to avoid delays in the sanctioning of loans and to ensure adequacy of financing.

Group Approach: SGSY also focus on the Group approach which involves organisation of the poor into Self Help Groups (SHGs) and their capacity building. Efforts are made to involve women members in each SHG. Besides, exclusive women groups continue to be formed. At the block level, at least half of the groups are exclusively women groups. SGSY particularly focus on the vulnerable groups among the rural poor. Accordingly, it has been decided that the SC/STs should account for at least 50% of the Swarojgaris, women for 40% and the disabled for 3%.

Despite planned efforts made over the past few decades, rural poverty in India continues to be worrying. While the anti-poverty programmes have been strengthened in the successive years during the course of the period by launching various programmes like IRDP, TRYSEM, DWCRA, SITRA, resulting in the reduction of poverty levels in percentage terms from 54.9 percent of India's population in 1973-74 to 36.0 percent in 1993-94 and to 26.1 percent in 1999-2000, the number of rural poor has not reduced significantly.

To redress the situation at a faster pace, a new restructured self-employment programme, known as "Swarnjayanti Gram Swarozgar Yojana" (SGSY), has been launched from April 1999, by doing away with the anomalies in the multiplicity of earlier programmes, to focus pointedly on the issue of ensuring sustainable income generation among the assisted poor families to bring them above the poverty line. SGSY is to focus on the vulnerable sections of the society. Accordingly, SCs/STs will account for at least 50 percent women 40 percent and disabled 3 percent of those assisted.

Success of the concept of SHG had gained wide currency. Significant growth of SHGs as well as SHG-bank linkage, bankers' recognition of SHGs as a medium of rural business expansion as well as their acceptance of the peer pressure within the SHGs as a substitute for collateral securities, and the interest as well as confidence exhibited by the rural poor in the concept of SHGs for their economic well being together with their far better loan repayment behaviour made the development planners believe that there is a tremendous potential within the poor to help themselves and that microfinance through Self-Help Groups together with an element of additional financial support including technology infrastructure and marketing from the Government can be a better alternative to the existing methods of addressing rural poverty.

In the light of the above, SGSY, the holistic programme covering all the aspects of self- employment, made it obligatory that the objective of the SGSY will be achieved, interalia, by organizing the rural poor into Self-Help Groups (SHGs) through a process of social mobilization, enabling them to build their own organizations in which they could participate fully and directly and take decisions on all the issues concerning eradication of their poverty.

1.3.3 National Rural Livelihood Mission (NRLM)

National Rural Livelihood Mission (NRLM) launched in 2011 is a poverty alleviation project implemented by Ministry of Rural Development, Government of India. This scheme is focused on promoting self-employment and organization of rural poor. The basic idea behind this programme is to organize the poor into SHGs and make them capable for self-employment. In 1999 after restructuring Integrated Rural Development Programme (IRDP), Ministry of Rural Development (MoRD) launched Swarnajayanti Grameen Swarojgar Yojana (SGSY) to focus on promoting self-employment among rural poor. SGSY is now remodeled to form NRLM thereby plugging the shortfalls of SGSY programme. This is one of the world's largest initiatives to improve the livelihood of poor. This programme is supported by World Bank with a credit of $1 Billion. The core belief of National Rural Livelihood Mission (NRLM) is that the poor have innate capabilities and a strong desire to come out of poverty. They are entrepreneurial, an essential coping mechanism to survive under conditions of poverty. The challenge is to unleash their capabilities to generate meaningful livelihoods and enable them to come out of poverty.

1.3.4 Integrated Watershed Management Programme (IWMP)

IWMP is a centrally sponsored scheme under the Ministry of Land Resources, Department of Rural Development, Government of India. All the watershed development programmes like IWDP, *Hariyali*, NWDPRA *etc.* are now under one watershed development programme *viz.* IWMP, following the New Common Guidelines published by Government of India in 2008 in order to have a unified perspective by all stakeholders. The key features of common guidelines include innovativeness in the approach, delegation of powers, strengthening dedicated institutions, social, gender and economic equity in sharing enhanced productivity and livelihood, multitier ridge to valley system approach and centrality of community participation.

The IWMP is a holistic project with all essential components such as capacity building, livelihood activities, production system, natural resource management, and a dedicated institutional system for effective and comprehensive implementation. Under livelihood activities of IWMP, livelihood action plan is implemented through SHGs and/or their federation. Livelihood activities can be carried out either through the existing SHGs having good performance or new SHGs formed with a group of 5-20 persons. SHGs selected for implementing livelihood action plan should be homogeneous in-terms of their existing livelihood capitals, common interest and need. The beneficiaries should be marginalized communities, including SC/ST, landless/assetless people, women, *etc.* SHG does not have more than one member from a household. Priority is given to women SHGs. Each SHG shall make an application for financial assistance to the Watershed committee (WC). The initial amount up to Rs. 25,000 may be given as seed money to a SHG as the revolving fund after their proposed activity(s) has been approved by the WC. The SHGs will return the seed money on monthly basis in a maximum of 18 months. The SHGs may use the amount for a combined activity and/ or shall provide the above amount to the concerned members as individual loan against a specific activity for improving income. In case of individual support under the SHGs, the individual will be accountable to the main SHGs for finances and performance.

1.3.5 Agricultural Technology Management Agency (ATMA)

ATMA, defined as a semi-autonomous decentralized participatory and market-driven extension model, represents a shift away from transferring technologies for major crops to diversifying output. The ATMA model is a central government initiative of the 2005–06: Support to State Extension Programmes for Extension Reforms (SEPER) scheme, which was designed to be implemented by each state at the district level. The pilot test was initiated in 1998 under the Innovation in Technology Dissemination (ITD) component of the National Agricultural

Technology Project (NATP) with the support of the World Bank in 28 districts in seven Indian states. In 2005 the Government of India expanded the ATMA model to 252 districts under SEPER, and then in 2007 to all districts of the country. The Government of India has revised the scheme (Department of Agriculture and Cooperation, Ministry of Agriculture, Government of India, 2010).

ATMA represents a platform for integrating extension programs across line departments, such as animal husbandry, fisheries, and forestry; linking research and extension units in a district; and inviting farmer participation in decision making. Extension intervention is based on the Strategic Research and Extension Plan (SREP) prepared after a Participatory Rural Appraisal (PRA) in each district.

The Farm Information and Advisory Centre (FIAC) is the physical platform at the block level where farmers, members of the private sector, and extension field staff members from each line department meet to discuss, plan, and execute extension programs. The block technology team includes technical officers from various line departments and consults with the Farmer Advisory Committee (FAC), which includes the heads of Farmer Interest Groups (FIGs), at the FIAC to develop a block action plan. The block action plan is then approved for funding by the ATMA governing board.

The FIGs and SHGs can be formed by local NGOs and then organized into producer groups by extension staff. In the first year of operation, the capacity of the groups is built through awareness campaigns, exposure visits, and training courses.

Each state has a SAMETI (State Agricultural Management and Extension Training Institute), whose mandate is to strengthen the capacity of mid-level and frontline extension staff and orient them to the ATMA scheme.

1.4 Principles and Features of SHG Approach

The basic principles of the SHGs are group approach, mutual trust, organization of small and manageable groups, group cohesiveness, sprit of thrift, demand based lending, collateral free, women friendly loan, peer group pressure in repayment, skill training capacity building and empowerment (Lalitha and Nagaraja, 2002).

Self Help Groups are voluntarily formed informal groups. The members are encouraged to save on regular basis. They use the pooled resources to meet the credit needs of the group members. The groups are democratic in nature and collectively make decisions. Since the members are neighbors and have common interest, the group is a homogenous one and cohesiveness is one of the characteristic features of the group. Regular savings, periodic meetings, compulsory attendance, proper repayment and systematic training are the salient features of the SHG. Evidences from various developing countries throughout the world have shown that the poor can be helped by organizing themselves into SHGs.

According to the Planning Commission of India, a SHG is:

(i) A self-governed, peer-controlled, small and informal association of the poor, with an average size of 15-20 people, usually from socio-economically homogeneous families, organized around savings and credit activities.

(ii) Members of the SHGs meet weekly or monthly to discuss their common problems and share information to arrive at a solution.

(iii) Group members make efforts to rectify their economic and social problems through mutual assistance and encouraged to make voluntary monetary contributions on a regular basis.

Self Help Groups have the following advantages:

1. They encourage the poor to save. The poor become credit worthy and bankable customers and are not seen as beneficiaries. They reduce the transaction cost of lenders and borrowers.
2. Women are trained in new skills and technologies and the wage earning workers become micro entrepreneurs.
3. They help the poor to gain economic and social empowerment. Increased asset creation and savings, higher employment and improved social lives of members are the benefits to the members.

1.5 Dynamics and Effectiveness of SHG

To understand any SHG, characterization is very important. In this context, it is apt to mention the concept of group dynamics given by Kurt Lewin (1951). Kurt Lewin's (1951) field theory of group dynamics assumed that groups are more than the sum of their parts. The formula $B = f(P,E)$ summarizes this assumption. In a group context, this formula implies that the behavior (B) of group members is a function (f) of the interaction of their personal characteristics (P) with environmental factors (E), which include features of the group, the group members, and the situation. According to Lewin, whenever a group comes into existence, it becomes a unified system with emergent properties that cannot be fully understood by piecemeal examination. Lewin applied the Gestalt dictum, "The whole is greater than the sum of the parts", to groups.

Thus, characterization of the SHGs includes analyses of personal characteristics of group members, group features and the situational factors which influence the group dynamics, group effectiveness and group performance; therefore, the SHGs formed under different programmes need analyses on the basis of socio-personal, socio-economic and communication profile of the members of respective SHGs, features of group effectiveness such as participation, decision making, operation and management function, fund generation, group atmosphere, membership feeling, norms, empathy, interpersonal trust, social support, etc. and situational factors like financial assistance, capacity building, etc.

1.6 Microfinance

Microfinance (MF) has become, in recent years, a fulcrum for development initiates for the poor, particularly in the Third World Countries. The need for rural credit in India has been recognized even before independence by the erstwhile British Government as early as 1793 when it issued regulations for Taccavi loans to farmers. The cooperative credit societies act passed in 1904 provides necessary legislative support to the financing of agriculture. Till late fifties, cooperatives have been the major institutional source for all agricultural loans. The last four decades witnessed a rapid growth of banking network in India under the guidance and initiative of the Reserve Bank of India (RBI). In 1952 the All India Credit Survey Committee came out with the objectives of production and productivity, the stance of policy towards rural credit was to ensure provision of sufficient and timely credit at reasonable rates of interest to as large a segment of the rural population as possible. The strategy rested on three pillars: expansion of the institutional structure, directed lending to disadvantaged borrowers and at lower interest rates. The chosen institutional vehicles for the task were cooperatives, Commercial Banks and Regional Rural Banks (RRBs). Between 1950-69 emphasis was on the cooperatives, RRBs in 1970s, National Bank for Agriculture and Rural Development (NABARD) in 1980s and Local Area Banks in late 1990s. Financial deepening occurred but the development impact of rural finance was blunted. In 1991, that is, on the eve of reform, the rural credit delivery system was in poor shape.

India has long taken efforts to expand credit availability to rural areas. Early programs, which often yielded disappointing results, were gradually replaced by efforts to establish self-help groups (SHGs) and link them to banks. In 1992, India's National Bank for Agricultural and Rural Development (NABARD) piloted the concept with 500 groups. Since then, the SHG movement has witnessed tremendous growth that brought about one of the world's largest and fastest-growing networks for micro-finance (Deininger and Liu, 2009).

The SHG-led approach differs from traditional micro-finance in a number of ways. First, it does not exclusively focus on credit or savings but also includes emphasis on social empowerment, outreach, and capacity building. Recognizing that households' lack of human and social capital may prevent them from making good use of financial resources even if they had access to them, program organizers put a strong focus on encouraging the groups to establish regular meetings among group members and group savings. There is also an emphasis on outreach whereby existing groups are encouraged to help the "leftover poor" in their village to form SHGs. Second, the goal is not to establish a separate micro-finance institution but to use the group to intermediate in dealings with the formal sector and help households to create a "credit history" that will eventually allow them to access regular sources of finance. Finally, federation of SHGs is a central element not only with respect to peer monitoring and diversification of risks on the financial side but federations at village and higher levels are also used to assist in implementation of government programs, help SHGs provide other services- from technical assistance to marketing- and allow members' participation in local government (Deininger and Liu, 2009).

The SHG model in India combines savings generation and micro-lending with social mobilization. A typical SHG consists of 10-20 members who meet regularly to discuss social issues and activities and, during these meetings, deposit a small thrift payment into a joint bank account. Once enough savings have been accumulated, group members can apply for internal loans that draw on accumulated savings at an interest rate to be determined by the group. Having established a record of internal saving and repayment, the group can become eligible for loans through a commercial bank, normally at a fixed ratio (normally starting at 4:1) to its equity capital. Rules adopted by the group specify the periodicity of meetings, the amount to be saved per meeting, the length of repayment period, the interest rate to be charged on internal and external loans which can be higher than that at which the loans are received, as well as the amounts and mechanism by which loans are allocated.

Among the total cultivar households of 89.35 million the coverage by formal sources banks, microfinance institutions Self Help Groups (SHGs) is only 27 per cent and this coverage shows a distinct bias towards households with larger farm holdings. The stance of policy and effort has been to find ways and means to increase the flow of institutional credit to the indebted poor, reduce the procedural and documentation hassles which characterise lending to such poor and ensure that affordable credit is reached to them at the appropriate time in adequate measure. If the categories of the poor have to be financed, then there is an obligation to create opportunities in which they can use the credit in a meaningful way and it can be best done by creation of production and employment opportunities in the real sector through public investment. Among the variety of interventions in reaching the excluded is the SHG bank linkage model of NABARD, is an outstanding example of an innovation leveraging on community based structures and existing banking institutions.

The task force on microfinance defined microfinance as "Provision of thrift, credit and other financial services and products of very small amounts to the poor in rural, semi urban or urban areas for enabling them to raise their income levels and improve living standards". A large number of institutions, both in the formal and non- formal sectors, are today providing a variety of financial services using different delivery mechanisms. The task force defines microfinance institutions (MFIs) as those which provide thrift, credit and other financial services and products of very small amounts mainly to the poor in rural, semi-urban areas for enabling them to raise their income level and improve living standards". The institutions like NGOs, federations of SHGs, Mutually Aided Cooperative Societies (MACS), State and National cooperatives which provide specified financial services targeted to the poor, may be classified as Microfinance Institutions (MFIs).

Micro financing or group lending is being looked upon as the instrument that can be considered as the golden stick for poverty alleviation vis-à-vis rural development. International initiative Mohammed Yunus, popularly known as father of micro-credit system, started a research project in Bangladesh in 1979 and came out with ideas of micro-credit and that resulted in the establishment of Grameen Bank in 1983. In 1984, the federal minister of economic cooperation and the agency of technical cooperation of the Federal Republic of Germany undertook series of studies that resulted in a new policy of self-help group (SHGs) as financial intermediation between rural poor and financial institutions. In 1986, the participation of Asia and Pacific Regional Agriculturist Credit Association (APRACA) decided on a coordinated programme for promotion of linkage between banks and SHGs for rural savings mobilization and credit delivery to the rural poor. In 1989, the central bank of Indonesia with the involvement of Self Help Promoting Institution (SHPI) started a pilot project entitled "Linking banks and SHGs". In 1995 the World Summit for social development was held at Copenhagen. It emphasized the easy access to credit for small producers, landless farmers and other low-income individuals particularly women. In 1997, the world micro credit summit in Washington announced a global target of ensuring delivery of credit to 100 million of the world's poorest families, especially the women of those families by 2005.

In India, the first effort was taken up by NABARD in 1986-87 when it was supported and funded as an action research project on "Saving and credit management of self help groups" of Mysore Resettlement and Development agency MYRADA. In 1991-92 NABARD launched a pilot project to provide micro credit by linking SHGS with bank. In 1999, Reserve bank of India (RBI) had setup a micro-credit cell to make it easier to micro-credit providers to pursue institutional development process.

SHGs are a viable organized setup to disburse micro credit to the rural women for the purpose of making them enterprising women and encouraging them to enter into entrepreneurial activities. Of the total SHGs formed, more than 6 million have been linked with 35,294 bank branches of 560 banks in 563 districts across 30 states of the Indian Union. Cumulatively, they have so far accessed credit of Rs.6.86 billion. About 120 million poor, of which around a third belonging in the SC/ST category, have gained access to the formal banking system through the programme. It has reduced the incidence of poverty through increase in income and helped the poor to build assets and thereby reduce their vulnerability. It has enabled the households that have access to it to spend more on education than non client households. It has empowered women by enhancing their contribution to household income. Increasing the value of their assets by giving them better control over decisions affects their lives. It has contributed to a reduced dependency on informal moneylenders and other non-institutional sources. The SHGs empower women and train them to take active part in the socio-economic progress of the nation and make them sensitized, self-made and self disciplined. The SHG enhances the earning capacity of women between 20 to 30 per cent.

1.7 Livelihood and Livelihood Security of SHG Members

As the very purpose of SHG approach is to provide rural poor a better livelihood and livelihood security, the analyses on these fronts considering the members of SHGs formed under different programmes would reveal outcome of the SHG approach.

A livelihood comprises of people, their capabilities and means of living including food, income and assets. Tangible assets are resources and stores and intangible assets are claims and access. A livelihood is sustainable when it maintains or enhances the assets on which the livelihood depends. Many of the definitions of livelihood security currently in use are derived from the work of Chambers & Conway (1992). A livelihood comprises of the capabilities, assets (stores, resources, claims and access) and activities required for a means of living.

The idea of livelihood as defined above embodies three fundamental attributes *viz.* the possession of human capabilities (such as education, skills, health, and psychological orientation), access to tangible and intangible assets, and the existence of economic activities. People and their access to assets are at the heart of livelihoods approaches. In the Department for International Development (DFID) framework (1999), five categories of assets or capitals are identified, which are:

1. Human capital: skills, knowledge, health and ability to work
2. Social capital: social resources, including informal networks, membership of formalized groups and relationships of trust that facilitate co-operation
3. Natural capital: natural resources such as land, soil, water, forests and fisheries
4. Physical capital: basic infrastructure, such as roads, water & sanitation, schools, ICT; and producer goods, including tools and equipment
5. Financial capital: financial resources including savings, credit, and income from employment, trade and remittances.

Therefore, livelihood is the function of physical, social, financial, human and natural assets. The livelihood security refers to food and nutritional security, economic security, habitat security, educational security, social security and health security.

1.8 Empowerment of SHG Members

Empowerment of rural poor in general and rural women in particular is the key outcome of SHG movement in India since early 1990s. The word 'empowerment' means giving power. According to the International Encyclopedia (1999), power means having the capacity and the means to direct one's life towards desired social, political and economic goals or status. Empowerment provides a greater access to knowledge and resources, more autonomy in decision making, greater ability to plan lives, more control over the circumstances which influence lives, and freedom from customs, beliefs and practices. Thus, empowerment of rural poor is not just a goal in itself, but key to all global development goals. Empowerment is an active multidimensional process to enable rural poor to realize their identity and power in all spheres of life. This emphasizes the objective to study empowerment of the members of SHGs formed by various government and non-government agencies under different programmes. The extent of empowerment may be realized through certain indicators like self development, economic empowerment, social empowerment, political empowerment and information empowerment. Women's empowerment as gender parity also required to be analyzed with respect to production, resources, time, leadership and income.

1.9 Issues and Challenges to SHG-Bank Linkage

The SHG-bank Linkage Programme has its origins in a GTZ-sponsored project in Indonesia. Launched in 1992 in India, early results achieved by SHGs promoted by NGOs such as MYRADA, prompted NABARD to offer refinance to banks for collateral-free loans to groups, progressively up to four times the level of the group's savings deposits. SHGs thus "linked" became micro-banks able to access funds from the formal banking system. The linkage permitted the reduction of transaction costs of banks through the externalisation of costs of servicing individual loans and also ensuring their repayment through the peer pressure mechanism. The programme encompasses three broad models of linkage:

Model I: Bank - SHG - Members

In this model the bank itself promotes and nurtures the self-help groups until they reach maturity. It accounted for 16% of cumulative bank loan provided.

Model II: Bank - Facilitating Agency - SHG - Members

Here groups are formed and supported by NGOs or government agencies. The dominant model, it accounted for 75% of cumulative loans of banks.

Model III: Bank - NGO-MFI - SHG - Members

In this model NGOs act as both facilitators and MF intermediaries, and often federate SHGs into apex organisations to facilitate intergroup lending and larger access to funds. Cumulative bank loans through this channel were 9% of total.

Another model has been piloted recently by NABARD for facilitating the formation of SHGs for bank linkage in areas where there are no NGOs. This involves using the services of committed individual volunteers identified by bank.

In India, SHGs represents a unique approach to financial intermediation. The approach combines access to low-cost financial services with a process of self management and development for the women. SHGs are formed and supported, usually, by NGOs or Government Agencies. Linking the SHGs not only banks but also for wider development programmes. SHGs are seen to confer many benefits both economic and social and enable women to grow their savings and to access the credit along with formal banking services, active in village affairs, social campaigns, stand for local body elections, address social or community issues.

This pattern has now shifted with greater involvement of Government agencies in promotion of SHGs though it has also become less clearcut with Government agencies taking over NGOs and banks often promoting through local NGOs.

There are several instances of experiments made a positive impact on the income and employment situation of poor. Women from different social and economic levels are joining to SHGs. However, the barriers to entry for the poor are high not only do they have lower income but their incomes are usually more variable. To reduce these barriers for the poor means allowing varying deposit amounts and frequency perhaps with a specified annual minimum principles them to access credit and repayments again within specified norms.

Some of the issues and challenges are enlisted in below paragraphs (Ramakrishna *et al.*, 2013):

1. *Livelihood Promotion:* The basic problem in the rural poor is low level of standard of living especially in developing countries like India. After the intervention of SBLP, the standard of living of the rural poor has been increased and they can also get nutritious food for their lives.
2. *Monitoring System:* Under SBLP, the keen monitoring has been set up like involvement of bank officials, NGOs, Municipal Corporation, etc. to strengthen and make more strong the SBLP system. This resulted into the effective utilization of economic resources for the beneficial activities.
3. *Capacity Building:* Under bank linkages, both the SHG members and bank officials have to undergone with the capacity building process organized by the Government and the NABARD. It makes more efficient in operation and management of the system.
4. *Low Rate Bank Loans:* Under the SBLP, the intension is to provide the formal financial services to the rural poor at a affordable rate of interest. Simultaneously, with the involvement of the NABARD and the RBI this objective has been achieved and resulted into many families become economically strong to access the modern services.
5. *Micro Insurance Products:* After succession in the micro savings and micro credit, the system laid the foundation for micro insurance products. As the time passes and with the social requirement, the system introduced the micro insurance scheme to the members of SHGs for their beneficiary.
6. *Emergence of Federations:* Some of the matured and urban based SHGs have made some federations to get maximum benefits from the Government as well as financial institutions. Similarly, the NGOs and other institutions are putting efforts to form the federations involving rural and very poor SHGs.
7. *Technology for Financial Inclusion:* Due to the awareness of formal banking services among the members, the technological financial

innovations are laid the foundation to make financial inclusion among the SHGs families. This resulted in bringing of members into the mainstream.

8. *Functioning:* The SHGs are functioning with new looks like they have inculcated time management and heavy practice of cooperative principles etc made the healthier competition in the economy.

9. *Sustainability:* The SHGs are formed not only for a purpose or achieving any specified objectives. Therefore, they are established to sustain in the society for a long run in order to get maximum benefits for the empowerment and self reliance.

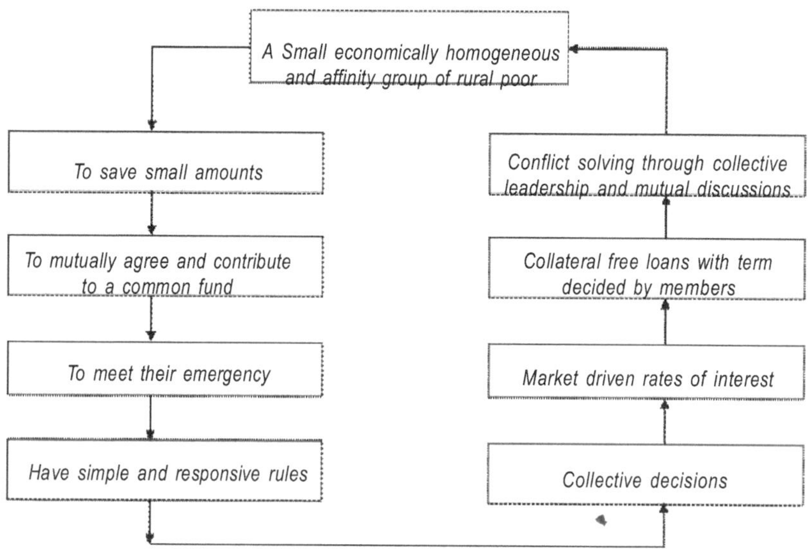

Fig. 1.2. Self Help Group (*Source*: Ramakrishna *et al.*, 2013)

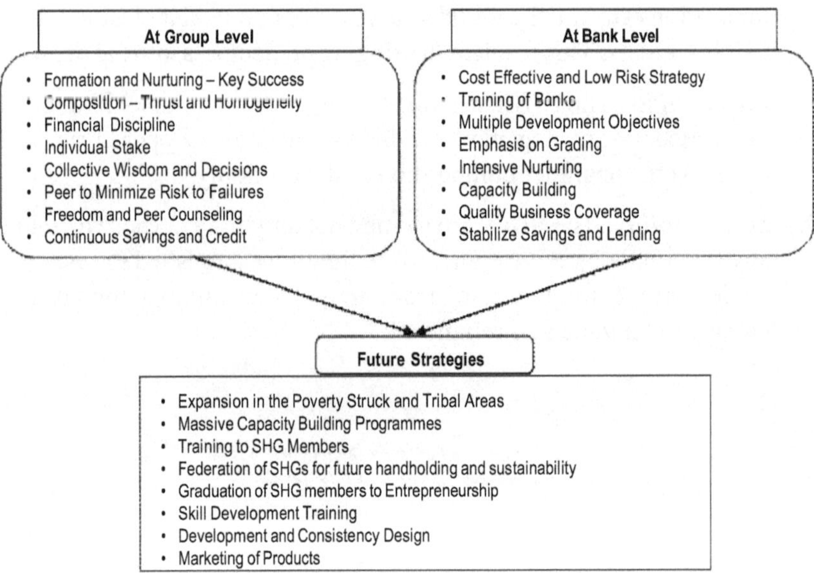

Fig. 1.3: Approach for SBLP Succession (*Source*: Ramakrishna *et al.*, 2013)

1.10 Conceptual Model

Based on the theoretical orientation, a Conceptual Model has been developed (Fig. 1.4).

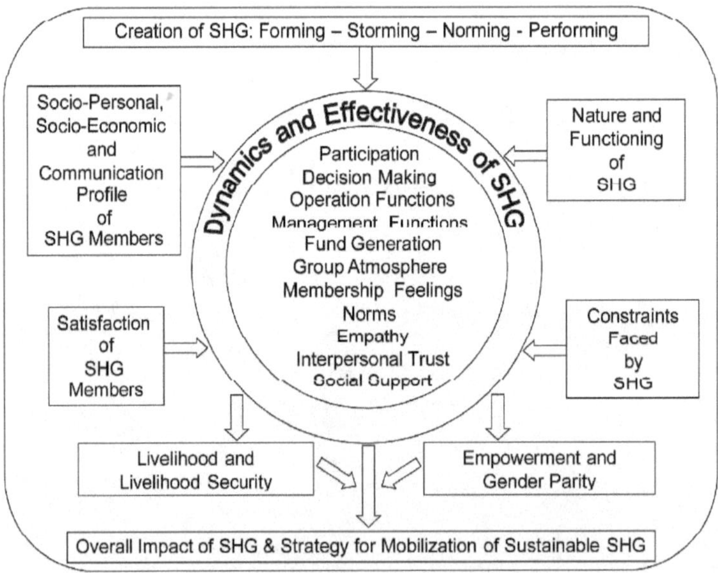

Fig. 1.4: Conceptual Model of the Study

Chapter 2

Glimpse of Past Research on SHG Approach

Self Help Group (SHG) and micro finance may be said a recent movement and the literature on SHG and micro finance is relatively one of the youngest. Review of past works in the field of SHG and related issues since 1995 has been considered, which helps to identify the research gap. Therefore, studies related to SHGs were reviewed and presented covering various aspects comprehensively.

2.1 Profile of the SHG Members

Hemalatha (1995) conducted study in Haryana and found that majority of women taking up pottery activity were illiterate, *Sirki* making women's education ranged from illiteracy to primary level and bakery unit were illiterate.

Kumaran (1997) conducted study in the rural areas of Tirupati block of Andhra Pradesh and reported that the mean age among the SHG members varied from 22 to 31 years. In another case study conducted by him on Self-Help Groups in Andhra Pradesh, it was reported that the average age among the members varies from 22 to 41 years. It was also mentioned that the 76 per cent of the members were illiterate and the remaining 24 per cent constitute literate who can read and write.

Prasad (1998) in the study conducted at Salem district of Tamil Nadu found that majority of women (60 per cent) were in age group of 25-45 years, of whom 40 per cent were in the age group of 30-40 years and 40 per cent were between 26-30 years.

Puhazhendhi and Jayaraman (1999) reported that 62 per cent of members in the informal groups belonged to the age group below 40 years.

Ganesamurthy *et al.* (2000) conducted a study of SHG in Erode and reported that 48.98 per cent of members belonged to the age group of 15-30 years and 40.82 per cent belonged to 30-45 years and 10.2 per cent belonged to 45-60 years. It was reported that 53.06 per cent of the members completed the school education and the remaining 46.94 per cent of them were illiterate.

Suriakanthi (2000) conducted research at Gandhigram of Dindigul district of Tamil Nadu and she found that 95 per cent of members and 75 per cent of the office bearers were illiterate.

Dwarakanath (2001) reported that 58 per cent of women involved in the production were in the age group of 19-35 years and more than 81 per cent of the members are in the middle age of 19-50 years.

Prita (2001) reported in the socio-economic profile of the members that 68.70 per cent were middle aged while 73.28 per cent were married. Illiterates comprised of 70.23 per cent. Nuclear families are documented with 87.02 per cent while large families with more than 4 members per family were marginally higher at 53.44 per cent.

Samar and Raman (2001) conducted a study of the SHGs in Tirupati (Andhra Pradesh) and found that 33 per cent of the members had no formal education, about 28 per cent had completed the elementary education, 18 per cent of them up to 10th standard, 17 per cent of members had reached to middle level and only about 4 per cent of the members had reached more than higher secondary level.

Sharada (2001) conducted a study on empowerment of rural women in SHGs in Prakasam district of Andhra Pradesh. She reported that 60 per cent of the members were very young.

Banerjee (2002) in his study conducted in Tamil Nadu reported that members in the age group of above 40 years participated actively in the group activities. Groups that were more than 3 years old had 42 per cent of the members of age above 40 years.

Raveendran (2002) reported that 63.3 per cent of group leaders in Tamil Nadu were illiterates where as in Kerala 58.3 percent got high school education.

Savitha (2004) conducted study in Mysore district and found that a higher percentage of women (39%) were illiterates and women with high school education were 21 per cent. It was reported that 73 per cent of the households had nuclear family while 27 per cent belonged to joint family.

Rao (2005) conducted study of women dairy co-operative societies which are involved in promoting SHGs. He reported that average age of respondents is 30 years. The respondents of all women dairy cooperatives had an average 2 years of formal education of which 50 per cent of them were illiterates.

Joseph and Easwaran (2006) reported that 53.85 percent SHG members in Mizoram, belonged to an age group of 40-60 years, followed by 30.77 per cent falling below 40 years and 15.38 per cent falling above 60 years. The mean age of members was found to be 48 years. It was reported that 51.28 per cent of the members belonged to secondary education followed by 33.33 per cent, with primary education and 7.69 per cent each of literate and higher secondary and above status. The mean year of education was computed to be 6.4 years.

Gangaiah (2006) reported that 67.3 per cent of the selected women members belonged to the age group of 26-40 years and 11.9 per cent of them to the age group upto 25 years.

Sharma *et al.* (2008) stated that raised literacy level could be helpful for the SHGs members to overcome cognition and to understand and gain required skill.

Meena and Singh (2012) reported that most of the SHG members in Eastern India were male (53%), aged from 27 to 47 years, had rural background (94%), experienced less than 6 year in SHG (61%), adopted agriculture as prime occupation (61%), educated up to primary level (51%) with low income.

According to Lokhande (2013), after joining the SHGs, the monthly average income and savings of the respondents had shown increase by 133.73 per cent and 76.27 per cent respectively. The confidence level of the respondents had increased to a great extent. They had reported active participation in socio economic activities. Overall personality index had shown growth of 59.01 per cent.

2.2 Functioning and Activities of the SHG

Puhazhendhi and Jayaraman (1999) reported that the SHG members taking up more than one activity increased from about 30 per cent during pre-group formation to about 53 per cent during post-group formation situation. They undertook supplementary activities such as animal husbandry, poultry *etc.* and non-farm activities like petty shop, *kirani* shop, flower selling business *etc.*

Dadhich (2001) reported that a large number of women had taken up subsidiary occupations like manufacturing of pickles, dairy, grocery shops and diversification of agricultural activities.

Dwaraknath (2001) reported that women had income generating opportunities through traditional and modern occupations such as spinning and weaving industry, agricultural activities, animal husbandry, hosiery, brass items, candle, cane items, carpets, *chappals*, chili powder, *khadi* / leather items, plastic items *etc*.

Puyalavannan (2001) conducted a study of CRUSADE (Center for Rural Systems and Development) an NGO in Thiruvalluvar district in Tamilnadu. He found that CRUSADE encouraged 100 members to take up income generating activities. Women members were granted loans up to 3000 for enterprises such as small business, goat rearing, fish vending, dairying, agricultural inputs *etc*.

Samar and Raman (2001) studied the working of SHGs in Tirupati of Andhra Pradesh and he found that dominant occupations of members are in maintaining Tiffin stalls, tailoring, selling vegetables, mike sacred thread making, *etc*.

Singh (2001) stated that the main activity of SHGs is to take up income generating activity and that each group is expected to concentrate on income generating activity.

Balakrishnan (2002) reported that the women under the micro credit schemes were engaged in small business activities such as making and sale of *idlies*, sale of flowers, vegetables and running of tea stalls.

Jha (2004) reported that there are *Mahila* self-help groups, *Stree Shakti* group and self-help money banks exclusively for women. The women are involved in rural development and entrepreneurships development. For example in vegetable and fruit preservation, mushroom cultivation, bee-keeping and production of cereal and millet- based agro-product as ready-made food, the role of women is much more than of the men.

Thejaswini *et al.* (2004) reported that the SHGs in Mysore district of Karnataka were practicing the enterprises such as tailoring, leaf plate making, pottery, *agarbatti* and bamboo works.

Tripathy (2004) reported that the most viable economic activities in the country were cotton coir rope making, coconut coir rope making, coconut leaf thatching, pickle manufacture, group leaf plate making, spices production, honey and food processing, *agarbatti* making *etc*.

Bharathi (2005) reported that majority of the SHG members were involved in entrepreneurial activities like vermicelli making (40.0%), *rava* making (20.8%), chilling pounding (16.7%) followed by tailoring, *papad* making *etc*.

Devalatha (2005) reported that the type of income generating activities taken up by the women were dairy, tailoring activities, *agarabatti* making, food item preparation, vermin composting, grocery shop, selling of clothes, goat rearing, *kirani* shop, candle making, bakery items *etc*.

Ganesan (2005) reported that the SHGs were running micro enterprises, which include paper cups and paper plates out of biomass, tailoring, screen-printing, mat making, sculptures and other ornamental activity like greeting cards, knitting plastic wire baskets making dolls and spoon using locally available coconut shells as raw material. They also prepare household products like pickles, *appalam*, phenyl, *etc.*

Kamaraju (2005) studied the SHGs in Thanjavur district which is an agrobased area and that the groups were collectively involved in milk farming, cattle farming, *appalam* making, *vadagam* and pickle making activities.

Anonymous (2006) reported that women of self help group of Kancheepuram district with help of Horticulture department started production of vegetables, jasmine and curry leaves.

Dasarathararamaiah (2006) studied the income generation activities through the Development of Women and Children in Rural areas (DWCRA) and found that Ram lamb rearing was undertaken by 8 per cent beneficiaries vegetable vending by 26 per cent, basket making 20.67 per cent, milk vending by 29.33 and 4.67 per cent petty trade, 1.33 per cent beneficiaries undertook leaves making and 10 per cent of beneficiaries undertook the fruit vending activity.

Gangaiah *et al.* (2006) studied the role of SHGs in the development of women in rural areas and reported that they were involved in activities like dairying, flower vending, tailoring, *idly* shop and cloth business.

SHG is a viable organized set up to disburse micro credit to the rural women and encouraging them to involve into entrepreneurial activities (Raheem, 2007).

2.3 Impact of SHG on Livelihood and Empowerment of Rural Poor

Srinivasan (1995) illustrates that realization of the poor women that they can take charge of their lives is a more significant gain of the SHGs. The confidence of the women that they can deal with external and modern institutions besides the feeling that they have competence to access and use finance and resources is another important gain. Further, she opines that access to financial resources is important for the development of poor women.

There is greater consensus however, on the role of microfinance in reducing Vulnerability. The provision of microfinance to SHG has been found to strengthen crisis-coping mechanisms, diversify income earning sources, build assets and improve the status of women (Hashemi *et al.*, 1996).

Applying an quasi-experimental design to 1991-92 data, Pitt and Khandker (1998), concluded that microcredit boosts household consumption, particularly when lent to women SHG. Finance against Poverty, bringing a new critical voice to the debate by showing the limitations of microfinance in bringing about poverty alleviation.

The failure of formal lending institutions and the apparent success of Bangladesh's Grameen Bank in reaching the rural poor have recently inspired numerous non-governmental organizations (NGOs) and governments of less developed countries to establish group-lending schemes to deliver credit at low cost and reasonable interest rates to small-scale rural entrepreneur (Coleman,1999). India is no exception in this case. From 1 April, 1999 Government of India initiated a programme called *Swarnjayanti Gram Swarojgar Yojana* (SGSY) which is a self-employment programme of Ministry of Rural Development that aims at providing assistance to the BPL rural poor for establishing micro-enterprises through bank credit and government subsidy to acquire an income-generating asset.

Coleman (1999) in a study considered the impact of group lending programmes in North East Thailand. This survey had shown that the impact of village banks that provide group-loans in villages is insignificant on physical assets, savings, productions, and productive expenditures and on other variables. However, it has positive impact on women's high interest debt because a number of members had fallen into vicious circle of debt from moneylenders in order to repay their loans on village banks. It has positive significant impact on women's lending out with interest because some members engaged in arbitrage, borrowing from village bank at low interest and then lending out money at mark up.

Puhazhendhi and Jayaraman (1999) in their study to document and evaluate the performance of informal groups in Chitradurga district of Karnataka and Periyar district of Tamil Nadu found that members taking up more than one activity increase from about 30 per cent during pre-group formation to 53 per cent during post group formation situation. They also reported that the average annual net income per member during pre-group formation ranged from Rs.6,763 to Rs.9,157 while the average net income per member during post group formation had ranged from Rs.10,531 and Rs.12,762. The increase in net incremental income was reported to be 68 per cent of new groups, whereas it was 100 per cent in stabilizing and stabilized groups. They also reported the impact of microfinance on social front. They observed that 95 per cent of group members were illiterate during pre-group formation and only 35 per cent of them were able to sign as result of group participation and 65 per cent of them were able to sign during post group formation. The literacy level of the family members also showed significant improvement where 55 per cent of them had graduated to the school level during post group formation stage. The result of the study reported that about 67 per cent of the sample household had effected improvements in their houses with additional rooms; whereas about 20 per cent of them improved sanitation facilities in their houses. About 3 per cent of the members had replaced the thatched grass roof to local tiles. Regarding consumption of food items, 32 per cent of them were able to afford vegetables and 30 per cent of the members were holding food stock to manage during the lean season. Radio was owned by five per cent of the members, which improved the perception of the members to a greater extent.

Puhazhendi (2000) describes the significant impact of SHGs on social empowerment, empowerment of women, employment generation, credit absorption, new income generating activities, savings pattern *etc*, in his study of SHGs in the state of Tamil Nadu. Further, he opines that the creation of income generating assets / activities through loans availed from banks has made significant impact on the overall economic status of the group members. He also points out that there is a positive impact of employment generation on 40 percent of the group members who had undertaken income generating activities. He observed that only 38 per cent of the members were able to sign during the pre-linkage period but as a result of group formation the literate members increased considerably and 85 per cent of them learnt to sign after the group formation. About 27 per cent of the members had educated their children upto the school level during the post linkage period. The study revealed that the members regularly started eating wheat and rice, after group activities which were earlier consumed by them during festivals. It is also observed that estimated average annual net family income of member during the post linkage period for all the groups was Rs.4,391, which was more than two times than that of the pre-linkage period. The estimated net incremental income was Rs.2,424 for all the groups and it was relatively more in good performance groups (Rs.2,967) than average and poor performance groups (Rs.1,650) and (Rs.1,299) respectively. He also reported that average ratio of Debt Service Liability (DSL) to Net Incremental Income (NII) worked out to be 0.60 and it was 0.53 and 0.81 respectively in good and average performance groups.

Puhazhendi, V and Satyasai, K.J.S (2000) conducted a study for NABARD on SHG-bank linkage programme. The study assessed the impact of microfinance on socio-economic conditions of 560 household members from 223 SHGs located in 11 states; Rajasthan (Northern region), Orissa and West Bengal (Eastern region), Madhya Pradesh and Utter Pradesh (Central region), Gujarat and Maharashtra (Western region), and Andhra Pradesh, Karnataka and Tamil Nadu (Southern region). They have observed that SHG–bank linkage programme has significantly contributed to the improvement in savings, assets, income levels and social conditions of the rural people.

Meyer (2001) observes that microfinance can contribute to poverty alleviation and food security. It does this through supplying loans, savings and other financial services that enhance investment, reduce the cost of self-insurance, and contribute to consumption smoothing. India has expanded microfinance, but it has not yet developed a strong system capable of serving massive numbers of poor in a sustainable fashion. Undoubtedly, the legacy of directed credit with its top-down approach to lending and the prevalence of highly subsidized state and national poverty projects and programmes retard the development of true market-oriented rural microfinance. The policy of supporting SHG linkages with banks has merit in a country with a large bank network, but it should not be the only model encouraged. Additional efforts are needed to create and nurture competitive MFIs willing to experiment with other models.

Dwarakanath (2001) reported that the DWCRA programme helped the rural women to earn an additional monthly income ranging from Rs.250-Rs.2000 depending on entrepreneurial activities taken up by them.

Samar and Raman (2001) reported that on an average, the SHGs have reported Rs.226 as income with maximum reaching Rs.3,314 for some SHG. Certain SHGs showed a loss in net income per member, the remaining SHGs registered positive net income per Member ranging from Rs. 12.90/- to Rs. 533.94/. To assess the impact of SHGs on income levels of members, two regression models were specified to find out the major determinants of a) SHG - net income per member, b) Average monthly income and found out that resources generated in current year, average educational levels, loan provided in current year, percentage share of SHG's expenditure in the total income of SHGs and age of SHGs showed expected signs.

It shows that there have been perceptible and wholesome changes in the living standards of SHG members in terms of ownership of assets, borrowing capacities, income generating activities, income levels and increase in savings. It indicates that the average annual saving per household registered an increase over three-fold (NABARD, 2002).

Puhazhendi and Badatya (2002) have made a study of impact assessment of the SHG bank linkage programme in India. The study compared the socio-economic conditions of 115 members in 60 SHGs during pre and post SHG situations to quantify the impact. Significant increase was observed in the mean annual savings, asset structure, average annual net income, average loan per member, overall repayment percentage and employment per sample households. Remarkable improvement was observed in social empowerment of SHG members in terms of self-confidence, better communication and involvement in decision-making, *etc.*

Manimekalai (2004) also remarked that the SHGs have the enough potential for establishing capacity building and self-efficiency among women.

Savitha (2004) conducted study on women empowerment decision making in agriculture by *Stree Shakthi* Groups in Mysore district and reported that the distribution of women according to social empowerment showed that majority had medium social empowerment and 26.67 per cent had high empowerment.

Experiences of Grameen bank in Bangladesh have shown that availability of collateral free tiny loans for income generating activities for poor have a significant impact on the lives of poor families (Yunus, 2004).

Ganesh (2005) reported that in Akola district of Maharashtra an SHG formed under SGSY in record time of one and half years, all the families belonging to BPL status have uplifted to 'Owner of Brick Kiln' status. Their net profits per 1000 bricks amount to Rs.550/- to Rs.650/- approximately. And their turnover has increased to more than Rs.3.5 lakhs.

Asokan (2005) reported that National Institute of Rural Development (NIRD) conducted a study on micro enterprises which are developed by SHGs in Kerala. The characteristics of micro entrepreneurs under SHGs revealed that a high proportion (90%) of them were unemployed prior to joining SHG and tailoring was found to be the most preferred activity (47%). The study also found that the average monthly turnover of micro projects taken by members of SHGs members was around Rs.1917 and net profit worked out to be Rs.700 per month. This indicates a high level of profit i.e. 60 per cent of individual units have investment less than Rs.5,000. A study conducted in Trichirapalli rural area found that before starting micro-enterprises their annual income was just 4,800 and after engaging in micro enterprises their annual income was increased to the tune of Rs.50,879.

Further, Khandker (2005) applying panel methods (using a 1999 resurvey) conclude that microcredit benefits the 'very poor' even more than the 'moderately poor'.

NABARD (2005) studied the impact of SHGs on economic empowerment of its member in Ballia district, Uttar Pradesh. And reported that there was an increase of the monthly income of each of the families by at least Rs.700/ month and this increase was solely due to the business that they were able to do by virtue of taking loan after the activities of SHG started.

Rao (2005) reported that the highest average annual household income (Rs.45,600) is from among respondents of *papads* and pickles and lowest (Rs.38,600) from respondents of chalk making activity. And the micro enterprises roughly provided 117- mandays/respondent which was a great contribution.

Dasaratharamaiah *et al.* (2006) reported that 10.0 per cent of beneficiaries had income between Rs.7,201 and above, 20.67 per cent have income between

Rs.4,801 to 7,200 and 31.33 per cent have income Rs. 3,601 to 4,800 and 38.00 per cent have income below Rs.3,600 per annum after implementation of DWCRA. And it was found that there are no persons without any income. And it was also found that 50 per cent of beneficiaries have less than 100 mandays of employment, 21.67 per cent of the beneficiaries have employment between 101 to 180 mandays, 20.00 per cent of the beneficiaries have employment between 181 to 240 mandays as against 8.33 per cent of the beneficiaries who have employment between 241 and above mandays of employment per annum.

Gangaiah *et al.* (2006) conducted study on impact of SHGs on income and employment generation. They reported that on an average the loans received generated 184 person days of employment per household. Non-farm activities generated higher number of person days of employment. Idly shop, cloth business and tailoring 300 each in number generated 240 person days of employment. They also found that SHGs had a favourable impact in generation of income in the village selected. The average income generated was Rs.19,578/-. Income generated in the selected activities shows that it varies from Rs.5000 per annum in case of idly shop to Rs.26,541 in the case of agriculture.

Joseph and Easwaran (2006) conducted a study to identify the constraints in functioning of SHGs and its impact on the members. And it was found that 51.28 per cent of respondents had income between Rs.25,000 to Rs.50,000. Majority of respondents had assets worth below Rs.1 lakh and more than one-half of the respondents as a whole (51.28%) had assets below Rs.1 lakh. They also studied the perceived impact of SHGs on Tribal development and found that majority of respondents reported high level of socio-economic impact of SHGs on tribal development. When studied the relationship between the composition and impact of SHGs. The perceived impact of SHG was found to be significantly associated with three variables duration of membership, members' participation and perceived group cohesion.

Dolli (2006) conducted a study on sustainability of natural resources management in watershed development project and found that the members of SHG had improved income (60%), self employment opportunities (66%), awareness (66%) and social contact (60%). Major benefits were expressed as awareness (53%), social contacts (53%), improved income (53%) and self employment opportunities (40%).

The existing literature on SHG- bank linkage programme reveals an overall picture of great promise on the socioeconomic well being of the members' households (Puhazhendi and Satyasai, 2000; Puhazhendi and Badatya, 2002; EDA Rural System and APMAS, 2006).

Suguna (2006) also remarked that the emergence of SHGs as silent revolution in the spread of rural credit for rural development.

Lolheihzovi (2007) considered SHGs as best engine of growth of human resource.

Roy (2007) undertaken quality assessment of SHGs in West Bengal and this was done by using twenty indicators like group meeting, members participation, group discipline, savings, micro-credit, financial management, economic and social initiatives and linkages with institutions. Factor analysis was done based on twelve indicators like cohesion in the group, initiative of the NGO, facilitation by PRI institutions, support of the bank, assistance of the line department, own initiative, livelihood opportunity, skill, infrastructure and market.

Recent literature has found a positive correlation between access to finance, economic growth and poverty alleviation (World Bank, 2008).

A very recent study in India by NCEAR (2008), found 25.3 percentage points net reduction in poverty of the households who were living below the poverty line, a significant drop from 58.3 per cent at the base level to 33 per cent in 2006. The study found that SHG-Bank Linkage programme has influenced the consumption pattern of member households.

Deininger and Liu (2009) reported SHG participation had significant economic impacts. Benefits significantly exceed program costs. Interestingly, benefits were not confined to those who had been more affluent to start with; in fact there is significant asset accumulation among the poorest of the poor. To the extent that they participate in SHGs, the poorest seem to be able to benefit not only socially but also economically.

Kumar (2010) while comparing the differences in quality of SHGs between SHGs under the umbrella of federations and other SHGs which are not part of federation observes that federation type SHGs are functioning well. He assesses the quality of SHGs by using NABARD CRI and also advised all banks to access the quality of SHGs using the CRI before every credit linkage.

Sahu (2010) assessed the quality of SHG in Northwest India based on the 13 indicators of the 200 SHGs, 27 per cent of the groups are found to be strong and stable; 10.5 percent groups are found to be poor rest being moderately stable.

Makandar (2011) in a study at Karnataka stated that women have been actively participating in decision making process after becoming members of SHG in the areas consumption of house hold items, education of children and their marriage, number of children, family, planning, purchase and sale of property.

NABARD (2011) reported that 76 per cent of the women members were able to interact with officials and 28 per cent of the members were able to save in banks; the result were seen in decision making in household matter, sending children to school, changing undesirable habits of their spouse, participating in Gram Panchayat election, access to bank credit after joining SHG (98%) as compared to mere two per cent before joining, increase in income by undertaking income generating activities, *etc.*

SHGs are playing a major role in removing poverty in the rural India today. The group-based model of self-help is widely practiced for rural development, poverty alleviation and empowerment of women (Das, 2012).

Rajendran (2012) reviewed 53 major studies carried out in India to identify the major trends, which indicates that micro finance and Self Help groups, by and large contributed to the development of core poor in terms of economic well being, alleviating poverty and empowerment leading to over all development of rural poor.

Hussain and Zafar (2012) reported that the financial status of households had improved due to improvement in access across formal credit institutions, since SHGs are linked with banks. Access to credit has enabled women to undertake income generating activities.

Empowerment means moving from a position of enforced powerlessness to one of the power. There are various indicators that define women empowerment. These indicators are mobility, autonomy, decision making, ownership of household assets, freedom from domination in the family, political and legal awareness, participation in social and development activities, contribution to family expenditure, reproductive rights, exposure to information media and participation in development programmes (Banerjee and Dutta, 2014).

Sharma *et al.* (2014) assessed the extent of effectiveness of SHGs in improving livelihood security and gender empowerment that took into account both male and female SHGs. SHGs enabled its members to grow their savings and to access the credit. SHGs also act as community platforms from which women become active in village affairs, take part in political decision making process at village level or take action to address social or community issues. It is also envisaged that involvement in SHG enabled its members to gain greater control over resources like material possession, knowledge, information, ideas and decision making in home, community, society and nation. Improved awareness of members about social issues is an important indicator of the capacity development of groups. The improved awareness levels enable group members to play a more effective role in community affairs and work towards achievement of common goals. SHG women are increasingly discussing and taking action on social and community problems like sanitation and pollution, management of village schools, various kinds of assistance to poor women, alcoholism, female feticide, dowry, access to drinking water, health, *etc.*

2.4 SHG and Microfinance

In the development paradigm, micro-finance has evolved as a need-based programme for empowerment and alleviation of poverty to the so far neglected target groups (women, poor, deprived *etc.*) and micro-finance has become one of the most effective interventions for economic empowerment of the poor. The experience across India and other countries has shown a robust potential of Microfinance to integrate with the development issues thereby significantly impacting the lives of poor. Rajendran (2012) has critically reviewed the various empirical studies carried out in India and it will help the researchers in the field of SHG and microfinance.

The National Bank for Agriculture and Rural Development (NABARD) introduced a pilot project commonly known as SHG linkage project in 1992.With a small beginning in 1992 as a pilot project, the active participation of Government, Banks, development agencies and NGOs has made the SHG movement as the world's largest microfinance programme. NABARD has defined micro finance as follows: "Micro finance is all about provision of thrift, credit and other financial services and products of very small amount to the poor in rural, semi urban and urban areas for enabling them to raise their standard of living". UN declared the year 2005 as year of micro credit since the policy makers of UN supported the view that micro finance is an instrument to fight against poverty. According to Nobel Committee, micro finance can help the people to break poverty, which in turn is seen as an important prerequisite to establish long last peace.

Out of the 53 research studies analysed by Rajendran (2012), majority of the (36) research studies focused on the impact of microfinance / SHG on women empowerment in various States and five research studies were conducted on the effect of microfinance / SHG on increase in income and access to financial sources and five studies were conducted to study the effect of microfinance on employment opportunity in non-form sector and only one study focused on poverty reduction with empowerment. Majority of the studies observed the positive effect of microfinance through SHG on economic, social, political and psychological empowerment, increase in income and employment opportunities, development of leadership qualities, enhanced participation in community activities and high degree of participation in domestic as well as in the society. Study conclusions all the 53 research studies are presented in tabular form for easy reference following Rajendran (2012).

Researchers	State	Salient findings of the study
Puhazhendi and Satyasai (2000)	Rajasthan, Orissa, West Bengal, Madhya Pradesh, Uttar Pradesh, Gujarat, Maharashtra, Andhra Pradesh, Karnataka and Tamil Nadu	The impact of microfinance was relatively more pronounced on social aspects than economic aspects.
Rao (2000)	Andhra Pradesh	SHGs showed a positive impact in respect of building of self-confidence, social development, skill formation and social empowerment.
Kallur (2001)	Karnataka	Group approach has brought many operative values like group support, thrift, group action and sustainability of women SHGs.
Manimekalai	andRajeswari (2001)	Tamil Nadu SHGs has helped the groups to achieve economic and social empowerment. It has developed a sense of leadership, organizational skill, management of various activities of a business, right from acquiring finance, identifying raw material, market and suitable diversification and modernization.
Nedumaran *et al.*(2001)	Tamil Nadu.	An increase in net income and social conditions of the members.
Puhazhendi and Badyata (2002)	Orissa, Chattisgarh andJarkhand	Women members income increased and increased opportunities of employment in non-farm and off farm employment in addition to social empowerment.
Krishnaiah (2003)	Andhra Pradesh	Women were able to diversify their activities by undertaking non-farm and animal husbandry related activities.
Satyasai (2003)	Andhra Pradesh,Tamil Nadu	Micro finance had positive impact in respect of self confidence, economic and social development and skill formation in Andhra Pradesh and social empowerment in Tamil Nadu.
Lalitha and Nagarajan (2004)	Tamil Nadu	Self Help Groups has laid the seeds for economic and social empowerment of women. Participation in group activities leads to changed self image, enhanced access to information and skills, broadened their knowledge about resource.

Rao (2004)	Karnataka, AP	Micro finance helped in improving the socio economic conditions of members.
Selvarajan and Elango (2004)	Tamil Nadu	Charging of high rate of interest is more oppressive causing hardships to the poverty stricken groups.
Tamizoli (2004)	Tamil Nadu	Possibility of employment near their homes and the concept of sisterhood is powerful and has changed women to sedentary from nomadic life.
Usha Rani *et al.* (2004)	Andhra Pradesh	Micro credit increased the access to financial resources and it made poor women financially self reliant.
Vadivoo and Sekar (2004)	Tamil Nadu.	Self Help Groups movement helped women collectively struggling against direct and indirect barriers to their self development and the social, political and economic participation.
Venkatachalam and Jeyapragash (2004)	Tamil Nadu	SHGs have made a silent revolution for the economic empowerment of poor rural women.
Rajivan (2005)	Andhra Pradesh	Enormous increase in self confidence among the women and significant reduction in dependence on money lenders and freedom from money lenders given them self respect.
Dhara and Nitra (2005)	West Bengal	Empowerment is only at elementary level and women are not aware of the banking procedure and leaders are finding it difficult to maintain account books.
Kabeer and Noponen (2005)	Jharkhand.	Members had more nutritious food, and enjoyed a favourable food situation and they had more of livestock, diversified cropping, high value crops, higher savings and reduced indebtedness.
Kumar (2005)	Haryana	Micro finance enhanced knowledge and skills of women.
Rajagopalan (2005)	Orissa	Women gained very significantly in terms of mobility, self confidence, access to financial services, building of own savings, competence in public affairs and improved status at home and in the community.

Joshi (2006)	Uttaranchal	Self Help Groups brought greater awareness regarding their roles, responsibilities and rights due to the participation in group meetings, training programmes and exposure visits have led to confidence building and social self esteem among women.
Moyle, Dollard and Biswas (2006)	Rajasthan	SHGs achieved both economic and personal empowerment in terms of collective efficiency, pro-active attitudes, self-esteem and self efficacy.
Sinha (2006)	Andhra Pradesh, Karnataka, Orissa and Rajasthan	Only 12 per cent SHGs taken issues on social justice such as domestic violence, dealing with dowry, prevention of child marriage, bigamy. Default rate was high at 28 per cent, 38 percent of very poor members have more over due, defunct groups emerging as an indicator of loan default.
Suguna (2006)	Andhra Pradesh	Improved social empowerment and capacity building of rural women.
Anjugam and Ramasamy (2007)	Tamil Nadu	The study has revealed that landless and marginal farm households and socially backward households participated more in the SHG-led microfinance programme.
Singh, Kaushal and Gautam (2007)	U.P	Group process had a positive significant relationship of empowerment and women's participation in Self Help Groups enabled them to gain self confidence, social and economic empowerment and capacity building.
Swain and Wallentin (2007)	Orissa, Tamil Nadu, Andhra Pradesh, Uttar Pradesh and Maharashtra	There is significant increase in the level of women empowerment over a period of time (2000-2003) and it does not mean that every woman has been empowered to the same degree, but on the average, the Self Help Group members were empowered over this period.
Kar (2008)	Odisha	Experience of SHGs in Orissa reveals that most of the groups are not able to do so purposively or compulsively. This aspect of the linkage programme has received little attention.
Oommen (2008)	Kerala	It is significant that the SC/STs have 'fairly improved' their ability to collectively bargain, to plan projects and to organise group activities besides improving their social position within their own groups and within the wider community. But there was poor economic empowerment measured in terms of improvement in assets and income.

Nirmala and Geetha (2009)	Kerala	Positive impact of microfinance. It contributes for improvement in household economic welfare and enterprise stability or growth and Micro finance is empowering women, bringing gender equality.
Raghavan (2009)	Kerala	By participating in various income generating cum developmental activities, the morale and confidence of women became very high. Capacity of the poor women of the State in several areas has gone up considerably. Status of women in families and community has also improved.
Banerjee (2009)	West Bengal	It was observed that from low-income group more people have shifted to high-income levels. This has reduced the inequality in the distribution of family monthly income.
Pillai andNadarajan (2010)	Tamil Nadu	The study concludes that microfinance has brought better psychological and social empowerment than economic empowerment.
Subramaniam (2010)	Tamil Nadu.	SHGs have ushered a silent revolution of poverty alleviation and women empowerment.
Makandar (2011)	Karnataka	Women have been actively participating in decision making process after becoming members of SHG in the areas consumption of house hold items, education of children and their marriage, number of children, family planning, purchase and sale of property.
NABARD (2011)	India	76 per cent of the women members were able to interact with officials and 28 per cent of the members were able to save in banks; the result were seen in decision making in household matter, sending children to school, changing undesirable habits of their spouse, participating in Gram Panchayat election. Access to bank credit after joining SHG (98 per cent) as compared to mere two per cent before joining, increase in income by undertaking income generating activities, etc.
Rajendran and Raya (2011)	Tamil Nadu.	There is a high level of political empowerment as compared to economic empowerment and poor level of social empowerment.

Reji (2011)	Kerala	Micro finance through groups haveempowered women in Kerala.
Sathiyabama and Saratha (2011)	Tamil Nadu	It was found that the qualities like democratic decision making, team spirit, team work, social mobility, self confidence, boldness to meet the officials, mutual help and in total the leadership qualities have improved to a significant level.
Surender, Kumari, and Sehrawat (2011)	Haryana	There is positive impact of SHGs on employment generation. Number of working days of beneficiaries in Livestock, Business and any others profession had increased after joining the SHGs. In this way, it is indicating that SHGs generate employment. Majority of beneficiaries accepted the improvement in economic condition after joining SHG.
Barua (2012)	Assam	The amount of loans provided to the members of SHGs were so small that it can't help the members to fight against poverty. There is the failure of SHGs, but not the failure of self-help.
Das (2012)	Assam	Observed that SHGs has a positive impact on women member and in many cases it is proved that SHG promotes empowerment SHGs have positive impact on decision making pattern.
Mohapatra (2012)	Odisha	SHGs contributed to socioeconomic empowerment of women at household level.
Sarkar & Baishya (2012)	Assam	Results suggest that women's access to credit has a role in improving the household decision making capacity, workforce participation rate and control over resources and even political and legal awareness, thereby opening/ opportunity for greater empowerment of women of Assam.

2.5 Constraints Faced by SHG

Hemalatha (1995) found that 50 per cent of women felt difficulties in repayment because of uncertain income though the group leaders did not agree with it.

Puhazhendi and Jayaraman (1999) attributed non-cooperation of individual members with group activities as well as personality clash between office bearers and group members resulted to the disintegration of groups. Lack of follow up action by the field staff of NGOs and also played a major role in disintegration.

Laksmi (2000) reported that the major constraints into effective and beneficial credit programming for women on a larger scale are lack of banking data disaggregating by gender and lack of an adequate analytical framework for integrating women into credit analysis. The policy concern for increasing access to credit and generating a sustained favorable impact on their socio-economic uplift can be transformed into practical reality only by achieving such integration.

Prita (2001) studied the performance of Self Help Groups in Dharwad district and found that the major constraints faced by the members were difficulties in diversification/starting of activities (41.67%), misunderstanding amongst SHG members (38.17%), lack of space for storage of materials (28.24%) and inadequate availability of raw material at the right time (16.03%).

Namboodiri and Shiyani (2001) reported that limited opportunity for income generating activities and loan portfolio dominated by consumption loan as major constraints.

Sharada (2001) reported that majority of women perceived the problem of lack of training and education i.e. 71.7 per cent.

Senthil and Sekar (2004) stated that political interference in the selection of beneficiaries under peoples plan, lack of timely credit facilities, lack of adequate credit, lack of adequate farm women oriented schemes and delay in operation of development programmes were the major constraints perceived by the SHG members.

Thejaswini *et al.* (2004) reported that majority of the respondents indicated that lack of training (85%), financial constraints (82%), poor quality of raw material (81%), high cost of production (77%), lack of quality aspects (73%), marketing problems (65%), lack of storage and warehousing facilities as the major constraints.

Darlingselvi (2005) reported that from the study conducted in Kanyakumari district that the members came across certain difficulties in marketing products in time.

Rao (2005) reported that though problems varied across activities, social taboos and lack of communication skills came out to be major factors. Lack of transportation, competition from established brands and lack of capital were voiced by women.

Reddy (2005) observes that the state of SHGs identifies key areas of weakness which undermine the sustainability of SHG movement. He identifies the major areas such as financial management, governance and human resource ranges from weak to average quality for a majority of SHGs.

APMAS (2006) on the SHG-Bank-linkage programme in India, addressed a wide range of issues including cases of dropouts from SHGs, internal politics, issues of social harmony and social justice, community actions, book-keepings, equity, defaults and recoveries *etc*., and sustainability of SHGs.

Joseph and Easwaran (2006) identified that lack of government attention was first and foremost problem (39% of members) followed by high rate of interest (33.43% of members), followed by insufficiency of loan for income generation, inability to repay the loan, *etc*.

Sharma (2007) stated that the SHG movement has not got success in some north-eastern states for reasons that are peculiar to the region. The study also observes banking constraints as a factor that hinders the quality of SHG in Northeast India.

The study by Upadhey and Kadam (2008) revealed that the SHGs were facing financial, administrative, marketing problems.

VOICE (2008) in a report submitted to Planning Commission mentioned that line department's cooperation was found to be low in majority of the states like Bihar, UP and Chhattisgarh. However, active cooperation was observed in states like Gujarat and Andhra Pradesh. The block level officials and extension workers were inadequately available.

Das (2012) observed that due to fast growing of the SHG-bank linkage programme in the State of Assam, the quality of SHG has come under stress. Some of the factors affecting the quality of SHGs are (i) the target oriented approach of the government preparing group, (ii) inadequate incentive to NGO's for nurturing their groups, (iii) lack of proper monitoring, (iv) absence of quality enhancement mechanism. The main problems faced by the SHGs are delay in sanctioning the loan (21.33%) followed by poor response of authorities (18.67%), lack of administrative experience (18%), difficulty to approach the authorities (12%), inadequate loan amount (14.67%), limited number of installment (8%) and the problem of lack of cooperation among the members (7.33%). Further, the SHGs are facing the problems of marketing, basic infrastructure, training and skill development avenues, lack of administrative experience in managing the affairs of the groups.

2.6 Research Gap and Context of Future Study

In India, as evident from the review of the literature, SHG movement revolves around Southern Indian States, where Governments and Non-Governmental Organizations are taking lead in its spread and implementation. Studies carried out in India, as evident from the above literature review, indicated that micro finance and SHG, by and large contributed to the development of core poor in terms of economic well being, alleviating poverty and empowerment leading to over all development of rural poor. However, the importance of SHG approach is more in the Eastern Indian States, which are poverty stricken. Moreover, there is dearth of study to analyze the SHGs formed by different agencies under different programmes to understand differential group dynamics, performance, effectiveness and impact, if any. Therefore, it is important to undertake study on performance of SHGs in such Eastern Indian States that would provide impetus for future SHG movement and associated issues.

Chapter 3

SHG Approach in Eastern Indian States: Experiences in Chhattisgarh

In India, majority of the people live in rural areas and are engaged in agriculture, earning a subsistence wage. Development which has been focused on them seems to have just passed by them. In India out of the 147.90 million rural households around 60 per cent are cultivation households. Eighty-eight per cent are headed by small/marginal farmers with holding less than 2 hectares (Throat, 2005). Therefore to finance this category of the poor, there is an obligation to create opportunities where credit can be used in a meaningful way.

Since Independence India has been changing in all fields including social systems but still a group of people still struggle to acquire equal rights in the society. All the so-called changes in the developing economy has not touched the major part of their lives as the social welfare programmes have not trickled down to certain parts of the society. It does not mean that the policy makers and the government have not drawn any welfare schemes but the schemes and programmes have not reached in whole for whom it was designed and hence it has lost its vigor and charm thus not bringing any benefits to the society for which it was planned and women sector is the most affected population of that society (Paramasivan, 2013).

In India, SHGs represents a unique approach to financial intermediation. The approach combines access to low-cost financial services with a process of self management and development for the rural poor including women. SHGs are formed and supported, usually, by NGOs or Government Agencies linking the SHGs not only banks but also for wider development programmes. This pattern has now shifted with greater involvement of Government agencies in promotion of SHGs. SHGs are seen to confer many benefits both economic and social and enable women to grow their savings and to access the credit along with formal banking services.

There is a massive mobilization of women taking place as a result of the SHG movement. The growth of SHGs incidentally has occurred during the economic reforms periods. The SHG movement has a good potential to serve both as a human face of the economic reforms as well as contribute towards women's emancipation. There is a major onus on all actors involved in SHGs promotion and development to further intensify their efforts in enabling SHGs to reach a mature stage with a major investment for capacity building of SHGs and proactive policies to help overcoming the constraints faced by SHGs to integrate them into the development programmes (Shyledra, 2008).

3.1. Statement of Problem

Though SHG movement is growing at a phenomenal pace and resulting in far reaching benefits to its members and also rural bank branches, it is facing a number of serious challenges. All these challenges could be summarized into two major challenges (APMAS & NABARD, 2009). These are uneven growth of SHGs in different parts and states of the country and uneven quality of SHGs across the country and issues related to their sustainability.

The success of SHG movement in South Indian States has augured well. However, the implementation of SHG approach is more demanding for rural poor of Eastern Indian States being the home of maximum below poverty line (BPL) families. Therefore, the effectiveness of SHGs in improving livelihood and empowerment of rural poor in Eastern Indian States hold paramount importance.

On this backdrop, a study was contemplated in one of the eastern Indian States i.e. Chhattisgarh with the following specific objectives:

1. To study the features of SHGs formed under different programmes and explore the profile of SHG members
2. To assess the dynamics and effectiveness of SHGs
3. To determine the impact of SHGs on livelihood of the members
4. To measure the influence of SHGs on empowerment of rural poor
5. To determine the overall impact of SHG approach and delineate the constraints faced by SHGs.

3.2. Scope and Importance

Many government and non-government organizations have undertaken the SHG approach under various schemes like NABARD's SHG-Bank Linkage Programme (SBLP), National Rural Livelihood Mission (NRLM), Integrated

Watershed Management Programme (IWMP), Agricultural Technology Management Agency (ATMA), etc. Therefore, present study would provide a scope to look into the differential effectiveness and functioning of SHGs formed under different programmes, which is important in providing impetus to the future planning on empowerment of rural poor through SHG approach, particularly in the state of Chhattisgarh.

The present study would reveal the characterization of different SHGs with reference to personal characteristics of group members, group features and the situational factors which influence the group dynamics, group effectiveness and group performance.

The impact of SHG on livelihood and livelihood security is perhaps most demanding and motivating factor; therefore, present study would also reflect the achievements made in this regard by different SHGs formed by various agencies under different programmes. Empowerment of rural poor in general and rural women in particular is the key outcome of SHG movement in India since mid 1990s. Whether the same result is evident in Chhattisgarh especially with reference to gender parity in production, resources, time, leadership and income would be surfaced through the present study.

The study provides a full insight into the constraints faced by the self help groups. Thus, present study is expected to reveal present scenario of self help group in Chhattisgarh.

Moreover, this study would assume great significance in creating database for realistic planning and implementation of future SHG movement. Findings will also serve as a bench mark for future evaluation. A study like this will help in adding to existing store-house of knowledge concerning SHG approach and related issues. It will also guide future researchers in deriving insight in understanding many aspects relevant particularly to SHG approach adopted under different schemes implemented by various government and non-government organizations.

3.3 Limitations of Study

Time was the major limitation both from the perspectives of the investigator and respondents during the course of interview-schedule survey. Some chance of biasness may be there as the information obtained as per the perceptions of respondents and existing situation. Universality and generalization cannot be claimed as the study was confined to a particular area and a small sample of SHGs and their members as respondents. The responses are mostly based on the memory and experience of the sampled respondents.

The study also considered official statistics available from different publications of the Govt. of India, Ministries of Government of Chhattisgarh, NABARD, etc. It is needless to refer to time lag in the publication of official statistics. However, the available latest statistics have been utilized.

The language barrier could not be completely avoided and an attempt was made to get the correct response by translating the interview schedule in rural dialect with the help of local people. Further, correctness of responses quite often depend on memory and despite best efforts of the investigator some margin would have been left for error to creep in.

Some scales, measurements and tests were used, but due to variations in rural conditions, the administration had to be done with minor modifications to ensure more reliable and effective results.

Inspite of the aforesaid limiting facts, sincere attempts and considerable thoughts have been exercised in making the study as objective and as systematic as possible.

Chapter 4

Research Settings to Study SHG Approach in Chhattisgarh

Present study was conducted in an Eastern Indian State 'Chhattisgarh', which was purposively selected keeping in view the importance of SHG movement in empowering rural poor in Eastern Indian States, home of the largest proportion of population below poverty line. To have in-depth micro level analyses, one district of Chhattisgarh was randomly selected, which happened to be the Kanker district.

4.1 Macro Setting: State of Chhattisgarh

The Snapshot of Chhattisgarh State which was formed in the year 2000 from erstwhile Madhya Pradesh is given below:

- Geographical area: 135192 square km.
- 27 districts with Raipur as the Capital city
- Major cities: Bhilai, Bilaspur, Raigarh, Durg, Korba
- Ranks 17 in terms of population and 10 in terms of area in the country
- GSDP growth consistently higher than National average. The State has clocked on an average more than 8 per cent growth rate over last 5 years
- GDP per Capita: INR 69,000 (FY2013-14)
- Established Exports Profile with exports aggregating INR 30 billion
- High reserves of Bauxite, Limestone, Quartzite
- Only Tin producing State of the Country
- 5 river basins, 3 major ones – Mahanadi, Indravati and Rihand.

4.2 Historical Perspective of Agriculture in Chhattisgarh

The State of Chhattisgarh, covering a geographical area of about 14 Mha, lies in the tropical belt of the eastern region of India between 17°43' to 24°50' N latitude and 80°15' to 84°20' E longitude. The map of Chhattisgarh is given in Figure 4.1. The state is bounded by as many as six states, Madhya Pradesh on its north-west, Uttar Pradesh on the north, Jharkhand on the north-east, Orissa on the east, Telangana on the south and Maharashtra on its west. The important rivers of the state are Mahanadi, Shivanand and Indravati. Mahanadi is the most important river of the state forming a Mahanadi valley, between the Vindhyas in the south. The Chhattisgarh state is predominantly is an agrarian one and famous for cultivation of rice and called rice-bowl of the country. About 80 per cent population depends on agriculture for its livelihood. Agriculture in the State is predominantly rainfed and the rainfall varies from 750 mm in northern part of the state to about 1600 mm in the Dandakaranya (Bastar) area of the state. The irrigation facilities are lagging behind in the state. Out of 14 Mha of available land about 35 per cent is used for cultivation. About 29 percent of the cultivated land is being irrigated. In irrigated areas, use-efficiency of applied irrigation water is very low (30 per cent or less). The state has got good scope of irrigation expansion and rainwater conservation. In general productivity of all the crops is low. The average productivity of rice and wheat varied from 1.65 to 1.60 t/ha, respectively. Cropping intensity varied from 101 (in Dantiwada district) to 139 (in Dhamtari) and overall cropping intensity of Chhattisgarh state is 137.

4.3 Climatic Potential and Constraints

The agro-climate of the State is characterized by transition between hot dry sub-humid to moist sub humid with dry summers and cool winters. The mean annual temperature varies from 25 to 28°C. The mean summer (April, May, June) temperature varies between 30 to 32°C and may rise to a maximum of 34 to 42°C in May. The mean winter (December, January, February) temperature varies from 17 to 19°C and may drop to a minimum of 8 to 13°C in January. The state receives mean annual rainfall varying between 1200-1600 mm, which covers more than 85 percent of annual demand of potential evapo-transpiration (PET) ranging between 1400-1600 mm. The south-western monsoon sets in the first week of June and extends till the first half of October. The peak period of the rainy season comprising end of June to September covers about 95 per cent of the total monsoon rainfall.

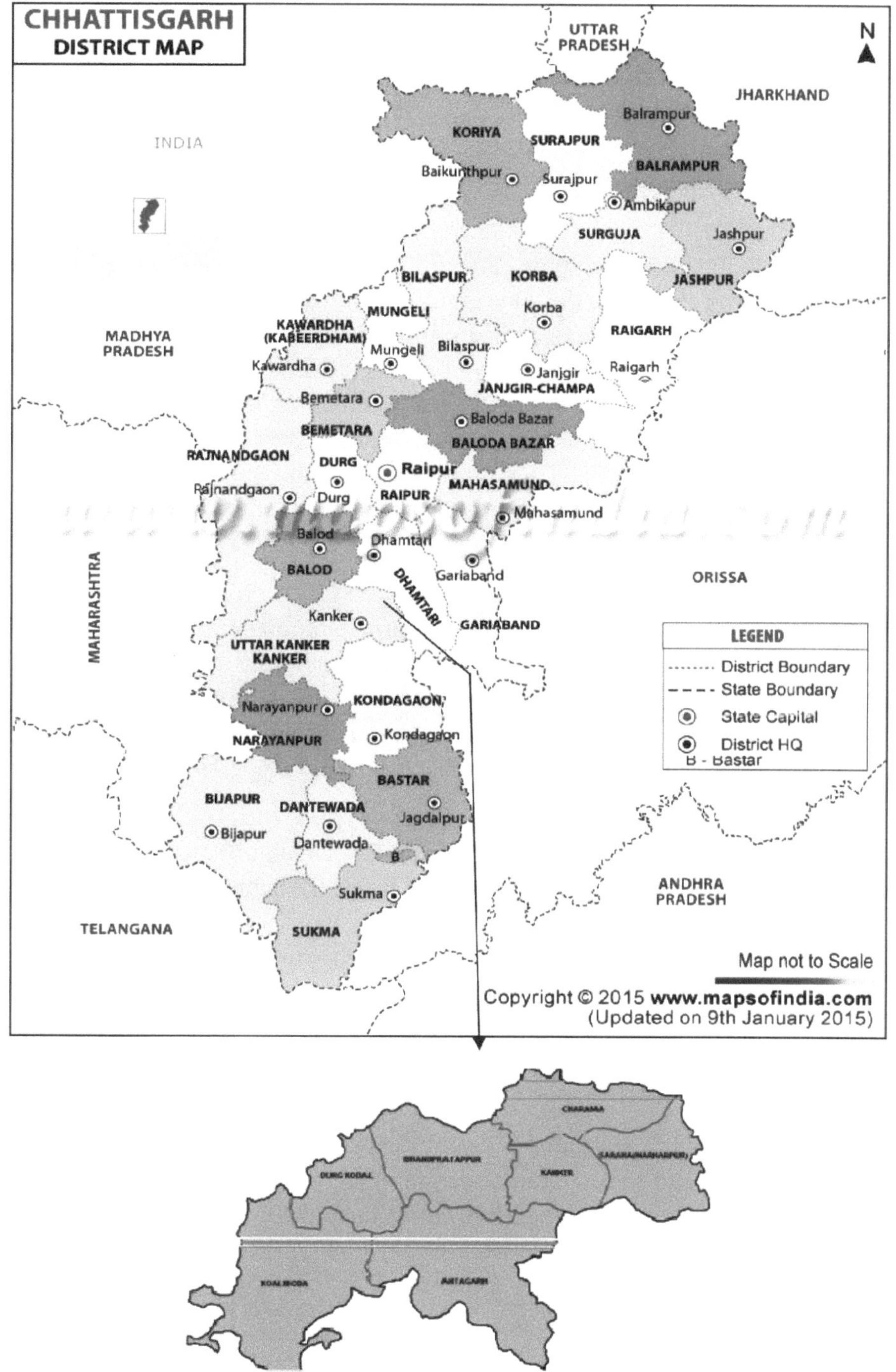

Fig. 4.1 Location of the Study (Kanker District of Chhattisgarh)

4.4 Agro-Climatic Zone of Chhattisgarh

On the basis of rainfall pattern, temperature, soil type and existing cropping pattern the Chhattisgarh state broadly divided into three major agro-climatic zones. These are i) North hill zone of Chhattisgarh, ii) Chhattisgarh plain zone and iii) Bastar plateau zone.

North hill zone of Chhattisgarh is situated between 20°-12' and 24°-6' N latitude and 80°-1' to 84°-8' E longitude. The altitude ranges from 293 m to 1200 m above m.s.l. The climate of the zone is sub-humid which is mild in summer and moderate in winter. The mean annual rainfall is about 1455 mm and about 87 per cent of annual rainfall is received during June to September. The total geographic area of the zone is about 7.26 Mha, which is 12.6 per cent of the state's area out of which about 45.2 per cent is under forests. The gross cultivated area is 2.82 Mha, which is 37.1 percent of the geographical area of the zone and only 4 per cent gross cropped area is under irrigation. Koriya, Surguja and Jashpur districts are coming under this zone.

Chhattisgarh plain zone receives rainfall between 1300 to 1600 mm. The rainfall is extremely erratic and the major part of precipitation occurs during July to August. The weather is warm and humid. The net irrigated area in the region is only 21.6 per cent of the net cultivated area. Canal system forms the major source of irrigation. The major crop of the region is rice. Raipur, Mahasamund, Dhamtari, Durg, Rajnandgaon, Kawardha, Bilaspur, Janjgir, Korba and Raigarh districts are coming under this zone.

Bastar plateau zone receives an annual rainfall between 1200 to 1600 mm with the maximum precipitation during July and August. About 62.8 percent area of this zone comes under forest. Net sown area is about 0.83 Mha. (21.4%) and only 13, 273 ha is under irrigation. The major crop of the region is rice but the productivity is very low less than 1 t/ha. Bastar, Kanker and Dantewada districts are coming under this zone.

4.5 Resource Poor People of Resource Rich Region

Though the State is rich in natural resources viz. soil, water (average rainfall is about 1430 mm) and forest, a large section of the people in the State are poor. At present about 37 % population live below the poverty line. Fifty per cent of rural population is considered as work force.

4.6 Micro Setting: Kanker District

The Kanker District is situated in the southern region of the state Chhattisgarh. Previously Kanker was a part of old Bastar district. But in 1998 Kanker got identity as an independent district of erstwhile Madhya Pradesh State. Kanker district covers an area of 7161 sq. km and lies between North latitudes 19°09' and 20°06' and East longitudes 80°30' to 81°15' with population of about 7.49 lakh. Small hilly pockets are seen throughout the area. Mainly five rivers flow through the district namely- Doodh river, Mahanadi, Hatkul river, Sindur river and Turu river. About 28.5 % of the geographical area is covered by forests. Groundwater development is about 10%.

At present the Kanker district has seven tehsils named Kanker, Charama, Narharpur, Bhanupratappur, Antagarh, Durgukondal and Pakhanjoor and seven blocks named Kanker, Charama, Bhanupratapur, Narharpur, Antagarh, Durgukondal and Koyalibeda. The total number of villages is 1004. The number of revenue villages is 995, whereas forest villages are nine.

The climate of the district is of Monsoon type. The May month is the hottest month and the December month is the coolest month. Average rain of the district is 1492 mm. 90% rainfall occurs during the June to October. Climate changes in to dry and wet. Dry climate is found in Kanker and Charama and the wet climate is found in Bhanupratappur. The district has a tropical climatic condition.

Red soil is found in most parts. Smooth and fertile soil is found in river valleys. Soil of this district can be divided into four sections, out of which Matasi type of soil is found at higher land surface than Kanhar and less than Bhata. This soil is appropriate for rice crop and is found in the maximum area of Kanker region.

The Economy of Kanker is based on agriculture and also it is a main job of tribes. The farmers who live in forest cut the trees before the rainy season and plough the land for agriculture and sow seeds of rice or other grain. This type of agriculture is called as Marhan or Dippa. After one and two years they prepare a new farm and leave the old one. And on the plain land the agriculture is done each and every year. People divide their farms by constructing the small partitions. Rice is the main crop of area but Wheat, Sugarcane, Gram,Green gram, Sesame, Maize are the other important crops. People also grow many types of vegetables. A lot of fruits like Mangoes, Bananas etc are also produced. The agricultural and socio-economic scenario of Kanker district as well as Chhattisgarh State as a whole is presented in Table 4.1.

Table 4.1: Socio-economic and agricultural scenario of Kanker district and Chhattisgarh State

Indicators	Kanker	Chhatisgarh
Total population	748941	25540196
SC/ST population (%)	56.40	43.37
Population density	116.42	154
BPL family out of total family (%)	35.15	39.93
Female work participation rate (%)	51.20	42.21
Agricultural labourer out of total workers (%)	24.14	31.94
Cultivators out of total workers (%)	60.50	44.54
Main workers out of total population (%)	38.13	32.26
Literacy rate (%)	70.30	71.04
Cultivated land out of total land (%)	41.31	35.89
Gross irrigated area out of gross sown area (%)	8.97	29.00
Cropping intensity (%)	107.52	137
Rice area (,00 ha)	1660.00	35400.00
Area under HYV of rice (,00 ha)	1.05	22.20
Average rice yield (kg/ha)	1950.00	2100.00
Food grain production (,000t)	311.60	7285.81
Per ha fertilizer consumption (kg)	55	79.40
Groundwater development (%)	9.22	24.65

Source: Directorate of Economics & Statistics, Govt. of Chhatisgarh; Agriculture Dept., Chhattisgarh

The SHG movement in Kanker district has been the initiatives of number of agencies as mentioned in the Table 4.2.

Table 4.2: The detail of SHGs formed by various agencies under different schemes in Kanker district of Chhattisgarh

Name of the Block in Kanker district	SHGs formed under different programmes				
	SBLP of NABARD	ATMA	NRLM	IWMP	Total
1. Antagarh	0	2	319	0	321
2. Bhanupratappur	20	6	261	38	325
3. Charama	100	14	606	69	789
4. Durgukondal	50	3	224	0	277
5. Kanker	70	20	555	37	682
6. Koylibeda	0	7	437	12	456
7. Narharpur	30	13	930	51	1024
Total SHG in Kanker district	270	65	3332	207	3874
Overall in Chhattisgarh	8739	314	33091	4523	46667

Source: NABARD, Directorate of Agriculture, *Aajeevika* (NRLM) and IWMP, Govt. of Chhattisgarh.

Chapter 5

Research Methodology to Study SHG Approach In Chhattisgarh

A research methodology is the logical and systematic planning and directing of a research process to accomplish research goals. The components of research methods used in the present study are as follows:

5.1 Hypotheses

5.2 Sampling plan

5.3 Variables and their measurements

5.4 The interview schedule

5.5 Statistical analyses

5.1 Hypotheses

The present study was undertaken with following hypotheses:

H_0 : There is no influence of members' profile on dynamics of SHG, no impact of SHG on livelihood and empowerment of members, and no difference between the SHGs formed under different programmes with respect to dynamics of SHGs, impact on livelihood and empowerment of rural poor/ SHG members.

H_1 : There is influence of members' profile on dynamics of SHG, impact of SHG on livelihood and empowerment of members, and difference between the SHGs formed under different programmes with respect to dynamics of SHGs, impact on livelihood and empowerment of rural poor/ SHG members.

5.2 Sampling Plan

The present study was conducted in the State of Chhattisgarh that was purposively selected for present study having relatively more percentage of rural households with at least one person belonging to a farmer-based organization and/or self-help group (Birner and Anderson, 2007). Kanker district was randomly selected out of 27 districts of Chhattisgarh for the present study. Out of seven blocks in Kanker district, one block i.e. Kanker block was randomly selected.

5.2.1 Selection of SHGs

Stratified random sampling method was followed for the selection of SHGs. As the SHGs have been formed by different agencies under different programmes such as NABARD's SHG-Bank Linkage programme (SBLP), National Rural Livelihood Mission (NRLM), Integrated Watershed Management Programme (IWMP), Agricultural Technology Management Agency (ATMA), *etc*, the clusters of SHGs under SBLP, NRLM, IWMP and ATMA were considered as four strata/ category of SHGs, thereby universe of the SHGs in this present study. Thereafter, three SHGs from each of the above mentioned four strata were selected following random sampling method. Thus, a total of 12 SHGs were considered as sample for present study inclusive of 10 SHGs of women and 2 SHGs of men. It is worth mentioning here that 90 per cent of SHGs are of women as reported by several studies in India.

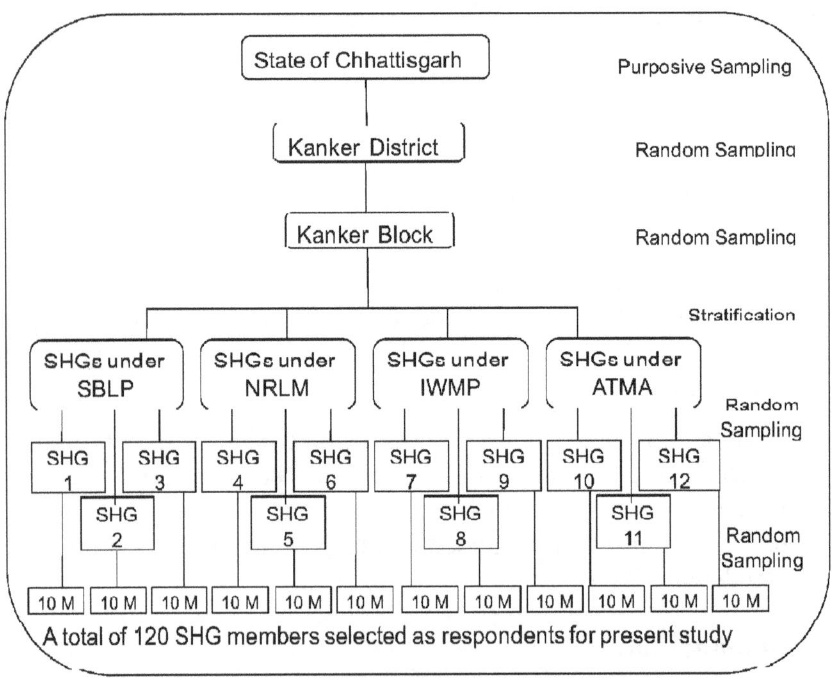

Fig. 5.1: Sampling Plan for Present Study

5.2.2 Selection of SHG Members as Respondents

SHGs used to have members of 10-20 persons; therefore, a sample of 10 members from each of the selected SHGs was chosen as respondents following random sampling technique. Thus, a total number of 120 persons being SHG members were surveyed for their responses on different aspects of SHGs.

5.3 Variables and Their Measurements

The variables were selected based on the objectives of present study. A brief detail of the variables considered in present study and their measurement procedures is given below.

5.3.1 Features of SHGs formed by various agencies under different schemes

Features of SHGs formed under NABARD's SHG-Bank Linkage programme (SBLP), National Rural Livelihood Mission (NRLM), Integrated Watershed Management Programme (IWMP), and Agricultural Technology Management Agency (ATMA) were explored through their basic information i.e. location, year of formation, membership detail, activities, financial detail of SHG, *etc.* The information obtained through secondary data as well as interview schedule survey of the selected respondents.

5.3.2 Profile of the SHG members

Socio-personal, socio-economic and communication profile of the sampled members of selected SHGs were considered. Socio-personal profile included age, sex, caste, education, family type and occupation of main earner of the family; while the socio-economic profile considered economic status, income and land holding. These variables were measured through nominal and ordinal scales.

In the present study, the age of the respondent was measured on the basis of three categories i.e. young (20-40 years), middle aged (41-60 years) and old (>60 years).

Sex of the respondents categorized as male and female.

The caste was measured on four categories i.e. general, other backward class (OBC), scheduled caste (SC) and scheduled tribe (ST).

Educational status of the respondent was measured on five categories i.e. illiterate, primary, secondary, higher secondary, graduate and above.

Family type was categorized as small (2-4 members), medium (5-7 members), large (8 members and above).

Occupation of the main earner of the SHG member's family was considered and it was classified as agriculture and allied, artisan, small business, labourer and others.

Economic status of the respondent categorized as below poverty line (BPL) and above poverty line (APL).

Average annual income of the SHG member was measured on five categories i.e. upto Rs. 25000, >Rs.25000 to Rs.50000, >Rs.50000 to Rs.75000, >Rs.75000 to Rs.100000, and >Rs.100000.

Total annual income of the family was measured by considering the total annual income of the family from all the sources. Accordingly, the respondents were categorized into the following five categories i.e. Upto Rs.50000, >Rs.50000 to Rs.100000, >Rs.100000 to Rs.150000, >Rs.150000 to Rs.200000, >Rs.200000.

In the present study, land holding of the respondent's family was measured on following categories *viz.* landless, marginal (<1 ha), small (1-2 ha), semi-medium (2-4 ha), medium (4-10 ha) and large (>10 ha).

Accordingly, the above-mentioned variables describing socio-personal and socio-economic profile of SHG member were included in an interview schedule developed for present study and responses of the sampled respondents were obtained. Thereafter, socio-personal and socio-economic profile of SHG members were analyzed and presented through frequency and percentage of respondents under defined categories of each of the variables.

Communication profile of the selected SHG member was assessed based on the use of communication/ information sources, which were categorized as personal localite, personal cosmopolite and mass media. Four sources under each of the above mentioned category were considered i.e. neighbours, friends, other farmers and village leader under personal localite source; *gram sevak/* village level worker (VLW)/ village agriculture worker (VAW), agricultural officers, irrigation department officials and other line department officers under personal cosmopolite source; and newspaper, magazine, radio and television under mass media/ impersonal cosmopolite source. These variables were measured through interval scale.

Depending on the frequency of use of each of these sources, responses obtained on a 5-point continuum scale *viz.* used twice or more in a week (5), used once in a week (4), used once in a fortnight (3), used once in a month (2) and used once in a while (1) through an interview schedule survey of sampled respondents.

The total score in case each of personal localite, personal cosmopolite and mass media sources use pattern of each SHG member was calculated. Thereafter, the mean score was derived separately for personal localite, personal cosmopolite and mass media sources by averaging scores of sampled members (10) of each selected SHG that might be varied between 0 (in case non use of any of the sources by all members) to 20 (in case of most frequently use of all the sources i.e. twice or more in a week).

5.3.3 Dynamics and effectiveness of SHGs

The dynamics of SHG has been operationally defined as the sum total of forces among the members of group based on certain sub-dimensions. Kurt Lewin popularized the term group dynamics to mean interaction of forces among group members in a social situation. It is the internal nature of the group as to how they are formed, what their structures and processes are, how they function and affect individual members, other groups and the organization. Pfeiffer and Jones (1972) identified indicators for analyzing group dynamics, which were participation, styles of influence, decision making procedure, task functions, maintenance functions, group atmosphere, membership feelings and norms. Vipinkumar and Singh (2001) mentioned few more dimensions influencing the group dynamics those are empathy, interpersonal trust and achievements of groups.

For the present study on dynamics of SHG, ten indicators selected by Ghosh *et al.* (2010) in formulation of group dynamics and effectiveness index for their study were used with minor modifications suiting to the need of present study. These were *viz.* participation, decision making, operation & management functions, interpersonal trust, fund generation, social support, group atmosphere, membership feelings, group norms and empathy. These variables were measured through interval scale.

The responses of sampled SHG members were obtained on each of five related statements under each indicator on a 3-point continuum scale based on degree of perception (2-always, 1-sometimes and 0-never for favourable statement and reverse score for unfavourable statement) with the help of an interview schedule.

The indicator-wise total score was calculated by summing the perceived scores of five statements for each SHG member. Thereafter, indicator-wise mean score was derived by averaging scores of sampled members (10) of each selected SHG that might be varied between 0 and 10. Overall dynamics of SHG was judged through the score obtained by adding mean scores of all 10 indicators which might be ranged from 0 to 100. The standard deviation values were calculated in derivation of mean scores showing the variations in perceptions of SHG members.

The effectiveness of SHG was assessed based on satisfaction of the SHG members having membership in SHG. The extent of satisfaction was expressed through perceptions of sampled SHG members on five issues *viz.* financial assistance, capacity building, empowerment, living condition and social status. These variables were measured with interval scale.

The responses of sampled SHG members were obtained on five statements corresponding to above-mentioned five issues on a 3-point continuum scale based on degree of agreement (2-agree, 1-undecided and 0-disagree) with the help of an interview schedule.

The issue-wise/ statement-wise mean perceived satisfaction score was calculated by averaging scores of sampled members (10) of each selected SHG. Overall satisfaction score was derived by summing the scores of five statements that might be varied between 0 and 10.

5.3.4 Impact of SHGs on livelihood of the members

Impact of SHG on livelihood of members was assessed from their livelihood security and level of living point of view.

The livelihood security refers to food and nutritional security, economic security, habitat security, educational security, social security and health security. Food & nutritional security refers to qualities of foods for the family members. Economic security considers sources of income, savings, loan/ credit availability, *etc.* Habitat security deals with condition of living, safety, insurance, adaptation at the time of any crisis/ emergency in family, *etc.* Educational security addresses to schooling of children, family members' literacy, communication ability, *etc.* Social security means recognition in society, membership in social organisations, *etc.* Health security indicates health condition, treatment of illness, readiness to meet health related expenditure, *etc.* Interval scale was used for the measurement.

Each of six indicators of livelihood security was included in an interview schedule developed for present study. During survey, the sampled respondents were asked to perceive each indicator on 5-point continuum i.e. very high (5), high (4), medium (3), low (2), very low (1) before and after joining the SHG.

The indicator-wise mean perception score was calculated by averaging scores of sampled members (10) of each selected SHG before and after joining the SHG that might be varied between 1 to 5. Overall livelihood security was derived by adding mean scores of all six indicators which might be ranged from 6 to 30.

Level of living is the function of physical, social, financial, human and natural assets. Physical assets refer to type of housing condition, sanitation, conveyance, electric, cooking, communication facility, *etc*. Social assets mean recognition, social and political participation, involvement in developmental works, common services used, *etc*. Financial assets involve sources of income, kinds of savings and investments, lending, borrowing, *etc*. Human assets indicate language competencies, education/literacy, management skill, mobility, *etc*. Natural assets consist of natural resources holdings viz. farm size, irrigated land, livestock, poultry, fishpond, *etc*.

Each of five indicators of level of living was assessed on the basis of perceptions of sampled respondents on 5-point continuum interval scale i.e. very high (5), high (4), medium (3), low (2), very low (1) before and after joining the SHG with the help of an interview schedule.

The mean perception score of each indicator both before and after joining SHG was calculated by averaging scores of sampled members (10) of each selected SHG, which might be ranged from 1 to 5. Overall level of living was measured through summation of mean scores of five indicators which might be varied between 5 and 25.

5.3.5 Influence of SHGs on empowerment of rural poor

Empowerment of rural poor through SHG approach was assessed with the help of following indicators *viz.* self development, social empowerment, economic empowerment, political empowerment and information empowerment in term of media exposure and awareness about development programmes. These variables were measured through interval scale.

The perceptions of sampled members of SHGs on each of these indicators were obtained on 5-point continuum i.e. very high (5), high (4), medium (3), low (2), very low (1) before and after joining the SHG during interview schedule survey.

The mean perception score of each indicator both before and after joining SHG was derived by averaging scores of sampled members (10) of each selected SHG. Overall empowerment was judged by summing the mean scores of five indicators which might be ranged from 5 to 25.

Women's empowerment from the gender parity point of view was evaluated with the help of 'women's empowerment in agriculture index' developed by IFPRI (2012). According to this index, the extent of empowerment was assessed out of 100% giving equal weight (20%) to five domains of empowerment *viz.* production, resources, time, leadership and income. The given weight (20%)

was further distributed over the sub-dimensions under each domain. The empowerment with respect to 'production' domain was assessed on the basis of two sub-dimensions i.e. input/opinion in productive decision (10%) and autonomy/ independency in production (10%). Similarly, 'resources' was having three sub-dimensions such as ownership of assets (6.66%), ability to purchase, sell and transfer of assets (6.67%), and access to credit (6.67%). Empowerment regarding 'time' domain was judged on the basis of two sub-dimensions - allocating own time for productive and domestic tasks (10%), and availability of time for leisure activities (10%). Empowerment in term of 'leadership' considered on two sub-dimensions i.e. membership of an economic/ social group (10%) and comfort in speaking in public (10%). The 'income' domain was evaluated in term of one sub-dimension - control over the use of income (20%).

The responses were obtained from the sampled members of each SHG in term of 'yes' or 'no' on each aforesaid sub-dimension. No score was given for 'no' response, while for 'yes' response, the score was given as per the equally distributed weight to sub-dimensions under each domain as described above. Thereafter, the total score of each domain was derived by summing the weight score of sub-dimensions under that domain. The overall empowerment score calculated by adding the scores of five domains, which might be ranged from 0 to 100%. Overall score of 80 per cent and above indicates the women's empowerment from the gender parity point of view as determined by IFPRI (2012).

5.3.6 Overall impact of SHG approach

The overall impact of SHG approach was measured considering 10 different issues which were: i) influence over economic resources and participation in economic decision making, ii) influence the individual development and confidence building, iii) influence in decision making of member in the household, iv) improved image of member in society and at home, v) increased capacity building of member, vi) increased awareness and knowledge of member, vii) increased participation and influence of member in social, community and other activities, viii) change in the attitude of the family members/ community regarding SHG member's empowerment, ix) increased mobility, development of networks and interactions with other members of his/ her group and community, x) maintaining economy and equity in income generating activity among the members. These were measured with interval scale.

The responses of sampled SHG members were obtained on ten statements corresponding to above-mentioned ten issues on a 3-point continuum scale (2-yes, 1-undecided and 0-no) with the help of an interview schedule.

The issue-wise/ statement-wise mean perceived satisfaction score was calculated by averaging scores of sampled members of selected SHGs. Overall impact score was derived by summing the mean scores of ten statements that might be varied between 0 and 20.

5.3.7 Constraints faced by SHGs

The constraints faced by each of the selected SHGs formed under certain programmes were listed as disclosed by the sampled members of each selected SHG.

5.4 Interview Schedule

In social science research, schedule plays a vital role. For the present investigation, interview schedule was prepared after completion of the review of literature and preliminary survey. The entire schedule was divided into six parts (see Appendix). Part A of the schedule contained general information on SHG. Part B included the variables describing socio-personal, socio-economic and communicational profile of the SHG member. Part C contained the statements corresponding to variables explaining dynamics and effectiveness of SHG to obtain response of SHG member on each of the statements. Part D consisted of the indicators of livelihood security and level of living for securing opinion of SHG member on each before and after joining the SHG. Part E covered the indicators of empowerment as well as five domains and sub-domains of women's empowerment in term of gender parity. Part F included the statements addressing the issues to assess overall impact of SHG. Open ended question to delineate constraints faced by SHG was included at the last part of the schedule.

5.4.1 Pre-testing of the schedule

The draft schedule containing various aspects of SHG and SHG member as described above was prepared and pre-tested in the non sampling area. The pre testing was done on the basis of responses obtained from 20 farmers five each from four non-sampling SHGs. The statements/ issues or aspects of the schedule were revised and/or modified in the light of feedback obtained through pre- testing of schedule. Efforts were made to keep the schedule as simple and appropriate as possible in order to make it easy and understandable. Opinions of experts were also taken into account in preparation of the final schedule.

5.4.2 Collection of data

The data were collected from the 10 sampled SHG members as respondents representing each of the 12 selected SHGs with the help of interview schedule prepared for present study. The data collection was carried out during February-

March 2015. After completion of data collection, data compilation and tabulation was done in the spread sheet using MS Excel programme. Finally, the data were analyzed.

5.5 Statistical Analyses

For analyzing the data for present investigation, the following statistical techniques were used based on the nature of data and relevance of information requirement according to the objectives of study.

5.5.1 Frequency and Percentage

These tools were employed to ascertain the distribution pattern of the respondents in different categories of variables considered to describe socio-personal and socio-economic profile.

5.5.2 Mean and Standard Deviation (SD)

Mean scores were calculated with respect to different indicators/ issues/ statements corresponding to respective parameters assessed such as communication/ information sources use pattern of SHG members, dynamics of SHG, effectiveness of SHG based on satisfaction of SHG members being part of SHG, livelihood security of SHG members, level of living of SHG members, influence of SHG on empowerment of SHG members/ rural poor and women's empowerment taking into consideration of gender parity.

The standard deviation values were calculated in derivation of mean scores on each occasion showing the variation in perceptions of SHG members.

5.5.3 Correlation and Regression Analyses

Correlation analyses between dynamics of SHG as dependent variable and independent variables like socio-personal, socio-economic, communicational characteristics of SHG members, satisfaction of SHG members being part of SHG were performed.

Subsequently, step-wise multiple regression analyses carried out considering dynamics of SHG as dependent variable and above-mentioned those independent variables, which showed significant correlation coefficient values.

In multiple regression analyses, all the variables included in the equations may not be equally contributing or have significant effect on the dependent variables. As such there is a need for dropping out the unimportant or non-significant variables from the equations. Step down regression technique is a method in which the insignificant variables are dropped from the analyses in step-wise manner; starting from the least significant variable. Ultimately, at the end of the analyses, the regression model develops the regression equation of the dependent variable with the independent variables having significant coefficient at the specified level of significance (used to consider upto 5% level).

A correlation coefficient matrix formulated taking into consideration dynamics of SHG, livelihood security of SHG members, level of living of SHG members and empowerment of SHG members.

5.5.4 Tests of Significance

t-test was performed to understand the significant difference (if any) between perceptions of SHG members representing the SHGs formed under a specific programme and that of SHGs formed under another programme with respect to each of parameters studied *viz.* dynamics of SHG, livelihood security of SHG members, level of living of SHG members, empowerment of SHG members and overall impact of SHG.

For the test of independence, also known as the test of homogeneity, 'Chi-square (χ^2) test' was conducted. The chi-squared statistic being at or larger than the 0.05 critical point) is commonly interpreted by researchers as justification for rejecting the null hypothesis.

Chapter 6

SHGs in Improving Livelihood and Empowerment of Rural Poor in Chhattisgarh

Findings of the present investigation on effectiveness of self-help groups (SHGs) in improving livelihood and empowerment of rural poor in Kanker district of Chhattisgarh are presented under the following headings:

6.1 Features of SHGs formed under different programmes

6.2 Socio-personal, socio-economic and communicational profile of the members of SHGs

6.3 Dynamics and effectiveness of SHGs

6.4 Impact of SHGs on livelihood of the members

6.5 Influence of SHGs on empowerment of rural poor

6.6 Overall impact of SHG approach

6.7 Constraints faced by SHGs

6.1 Features of SHGs Formed under Different Programmes

Brief detail of 12 SHGs, located across seven villages and seven GPs in Kanker block of Kanker district, Chhattisgarh, is presented in Table 6.1. Sample of three SHGs under NABARD's SHG-Bank Linkage Programme (SBLP), National Rural Livelihood Mission (NRLM), Integrated Watershed Management Programme (IWMP), Agricultural Technology Management Agency (ATMA) were studied.

SHG 1, SHG 2 and SHG 3 were formed in the year 2011 under NABARD's SBLP in village & GP Kanharpuri. Each group was having 10 female members with credit and thrift as the main activity. These groups also engaged in Fisheries (SHG 1), Agriculture and Poultry (SHG 2), and Agriculture and Goat keeping

(SHG 3). In case of SHG 1 and SHG 3, membership fee was Rs. 100 per month; having group's bank account at CBI (since 2011) with current balance Rs. 30000 and Rs. 24000, respectively; did not receive any loan. Membership fee in SHG 2 was Rs. 50 per month; bank account at CBI (since 2011) with balance Rs. 11000; no loan received.

SHG 4, SHG 5 and SHG 6 were formed under NRLM. SHG 4 and SHG 5 were formed in 2012 and 2007, respectively, in village & GP Nathiya Navagaon with 10 female members in each. SHG 6 was formed in 2009 in Nandanmara village, Arjuni GP with 10 female members. Credit and thrift (SHG 5 and SHG 6), mid-day meal in village school (SHG 4 and SHG 6), vegetable cultivation (SHG 4 and SHG 6), selling of agricultural inputs (SHG 4 and SHG 5) were the activities of these groups. In case of SHG 4, membership fee was Rs. 50 per month; having bank account at CGB (since 2012) with current balance Rs. 15000, no loan received. Membership fee was Rs. 50 per month in SHG 5 having its bank account at CGB (since 2007) with balance Rs. 54000, loan received @ Rs. 25000 thrice. In SHG 6, membership fee was kept as Rs. 100 per month, having group's bank account at CGB (since 2009) with balance Rs. 30000.

SHG 7, SHG 8 and SHG 9 were formed under IWMP. SHG 7 and SHG 9 were formed in 2009 and 2006 with 20 male and 11 female members, respectively, in village & GP Aturgaon. SHG 8 was formed in 2002 in village & GP Kulgaon with 11 female members. All three groups were engaged with fisheries activity. Besides, credit and thrift (SHG 7), agriculture (SHG 7 and SHG 9), mid-day meal in village school (SHG 8), cooperative ration shop (SHG 8), selling of fertilizers (SHG 9) were the different activities undertaken by these groups. In SHG 7, membership fee was Rs. 500 one time, having group's bank account at SBI (since 2009) with balance Rs. 5000 after providing credit of Rs. 60000 to group members, no loan was received. Membership fee was Rs. 20 per month in SHG 8, having group's bank account at SBI (since 2002) with current balance of Rs. 250000, no loan received. In case of SHG 9, membership fee was Rs. 20 per month, having group's bank account at SBI (since 2006) with balance Rs. 10000 after providing credit of Rs.150000 to members), loan of Rs. 200000 was received for fertilizer marketing.

SHG 10, SHG 11 and SHG 12 were formed under ATMA. SHG 10 was formed in 2011 in village & GP Kodagaon with 12 male members. SHG 11 and SHG 12 were formed in 2011 and 2012 with 10 and 11 female members, respectively, in village & GP Bardevari. All three groups were engaged with diversified activities. Agriculture production, goat keeping, and transporting agricultural commodities were found to be the activities of SHG 10. SHG 11 was engaged in mushroom production and vegetable production. Credit and thrift, jute production, and mid-

Table 6.1: Brief description of the SHGs selected in Kanker district of Chhattisgarh

Sl. No.	Name and address of SHG	Scheme	Year of formation	No. of members	Main activities
SHG 1	*Pratigya* SHG, Village: Kanharpuri, GP: Kanharpuri	SBLP of NABARD	2011	10 female	Credit and thrift, fisheries.
SHG 2	*Suhavna* SHG, Village: Kanharpuri, GP: Kanharpuri	SBLP of NABARD	2011	10 female	Credit and thrift, agriculture, poultry.
SHG 3	*Ahilya* SHG, Village: Kanharpuri, GP: Kanharpuri	SBLP of NABARD	2011	10 female	Credit and thrift, agriculture, goat keeping.
SHG 4	*Jai Laxmi* SHG, Vill: Nathiya navagaon, GP: Nathiya navagaon	NRLM	2012	10 female	Mid-day meal in village school, vegetable production, selling of agril. inputs.
SHG 5	*Jagriti* SHG, Vill.: Nathiya navagaon, GP: Nathiya navagaon	NRLM	2007	10 female	Credit and thrift, selling of agricultural inputs.
SHG 6	*Jai Laxmi* SHG, Vill.: Nandanmara, GP: Arjuni	NRLM	2009	10 female	Credit and thrift, mid-day meal in school, vegetable production.
SHG 7	*Jai Bhavani Seva Samiti* SHG, Village: Aturgaon, GP: Aturgaon	IWMP	2009	20 male	Credit and thrift, fisheries, agriculture.
SHG 8	*Bhavani* SHG, Village: Kulgaon, GP: Kulgaon	IWMP	2002	11 female	Cooperative / ration shop, mid-day meal of village school, fisheries.
SHG 9	*Swayam Shakti* SHG, Village: Aturgaon, GP: Aturgaon	IWMP	2006	11 female	Agricultural production, fisheries, fertilizer marketing.
SHG10	*Vikas* SHG; Village: Kodagoan, GP: Kodagoan	ATMA	2011	12 male	Agriculture production, goat keeping, transporting agril. commodities.
SHG11	*Danteshwari* SHG, Village: Bardevari, GP: Bardevari	ATMA	2011	10 female	Mushroom production, vegetable cultivation.
SHG12	*Tulsi* SHG, Village: Bardevari, GP: Bardevari	ATMA	2012	11 female	Credit and thrift, jute production, mid-day meal in village school.

day meal in village school were the activities of SHG 12. In SHG 10, membership fee was Rs. 100 per month, having group's bank account at SBI (since 2011) with current balance Rs. 60000, an amount of Rs. 50000 was received as loan from SBI. Membership fee was Rs. 20 per month in SHG 11, having group's bank account at Bank of India (since 2011) with current balance Rs. 30000, no loan received. In case of SHG 12, membership fee was Rs. 100 per month, having bank account at Bank of India (since 2012) with balance Rs. 43000 at present (March 2015).

The SHG 7 and SHG 10 were male SHGs formed under IWMP and ATMA, respectively, while rest 10 was female SHGs.

6.2 Socio-Personal, Socio-Economic and Communicational Profile of Members of SHGs

The characterization of the SHG members is important as it influences the group dynamics, group effectiveness and group performance; therefore, the characterization was done on the basis of socio-personal, socio-economic and communication profile of the sampled members of selected SHGs which were formed under SBLP of NABARD, NRLM, IWMP and ATMA.

6.2.1 Socio-Personal, Socio-Economic and Communicational Profile of Members of SHGs under NABARD's SBLP

The Table 6.2 presents the socio-personal and socio-economic profile of the members of three selected SHGs (SHG 1, SHG 2 and SHG 3) formed under NABARD's SBLP in Kanker block of Kanker district in Chhattisgarh.

Most of the members (90%) were young (20-40 years age), while three out of 30 members were middle aged (41-60 years).

All the members of selected three SHGs (SHG 1, SHG 2 and SHG 3) were female.

The members of all three selected SHGs represent OBC, SC and ST categories; however, it was overall dominated by SC category (53%). The representation of SC was highest in SHG 2 (90%) and that of ST was highest in SHG 1 (40%); while presence of both SC and ST categories were equal in SHG 3.

Educational qualifications of SHGs' members were varied. About 40 per cent members of SHG 1 were having secondary level of education. Six out of ten members of SHG 2 were having primary level education. In case of SHG 3, four members were illiterate. Thus, overall 37 per cent and 30 per cent of the members were having primary and secondary level of education, respectively. However, 27 per cent of the members were illiterate.

Table 6.2: Socio-personal and socio-economic profile of the members of selected SHGs under NABARD's SBLP

Characteristics	SHG 1 (n=10) Frequency (%)	SHG 2 (n=10) Frequency (%)	SHG 3 (n=10) Frequency (%)	Overall (N=30) Frequency (%)
1. Age:				
Young (20-40 years)	10 (100)	8 (80)	9 (90)	27(90.00)
Middle aged (41-60 years)	-	2(20)	1(10)	3 (10.00)
Old (> 60 years)	-	-	-	-
2. Sex:				
Male	-	-	-	-
Female	10 (100)	10 (100)	10 (100)	30 (100)
3. Caste:				
General	1 (10)	-	1 (10)	2 (6.67)
OBCSCST	2 (20)	-	1 (10)	3(10.00)
	3 (30)	9 (90)	4 (40)	16 (53.33)
	4 (40)	1(10)	4 (40)	9 (30.00)
4. Education:				
Illiterate	3 (30)	1 (10)	4 (40)	8 (26.67)
Primary	2 (20)	6 (60)	3 (30)	11 (36.67)
Secondary	4 (40)	2 (20)	3 (30)	9 (30.00)
Higher secondary	1 (10)	1 (10)	-	2 (6.66)
5. Family:				
Small (upto 4 members)	5 (50)	3 (30)	4 (40)	12 (40.00)
Medium (5-7 members)	2 (20)	5 (50)	5 (50)	12 (40.00)
Large (>7 members)	3 (30)	2 (20)	1 (10)	6 (20.00)
6. Occupation of main earner:				
Agriculture and allied	7 (70)	4 (40)	6 (60)	17 (56.67)
Artisan	1 (10)	-	-	1 (3.33)
Small business	1 (10)	3 (30)	1 (10)	5 (16.67)
Labourer	1 (10)	3 (30)	2 (20)	6 (20.00)
Others	-	-	1 (10)	1 (3.33)
7. Economic status:				
APL	3 (30)	2 (20)	1 (10)	6 (20.00)
BPL	7 (70)	8 (80)	9 (90)	24 (80.00)
8. Annual income of member:				
Upto Rs.25000	6(60)	7 (70)	4 (40)	17 (56.67)
>Rs.25000 to Rs.50000	4 (40)	1 (10)	6 (60)	11 (36.67)
>Rs.50000 to Rs.75000	-	-	-	-
>Rs.75000 to Rs.100000	-	1 (10)	-	1 (3.33)
>Rs.100000	-	1 (10)	-	1 (3.33)
9. Annual family income:				
Upto Rs.50000	2 (20)	4 (40)	4 (40)	10 (33.33)
>Rs.50000 to Rs.100000	5 (50)	2 (20)	4 (40)	11 (36.67)
>Rs.100000 to Rs.150000	1 (10)	-	2 (20)	3 (10.00)
>Rs.150000 to Rs.200000	1 (10)	1 (10)	-	2 (6.67)
>Rs.200000	1 (10)	3 (30)	-	4 (13.33)
10. Land holding of family:				
Landless	2 (20)	3 (30)	7 (70)	12 (40.00)
Marginal (<1 ha)	5 (50)	5 (50)	1 (10)	11 (36.67)
Small (1-2 ha)	2 (20)	-	1 (10)	3 (10.00)
Medium (4-10 ha)	1 (10)	1 (10)	1 (10)	3 (10.00)
Large (>10 ha)	-	1 (10)	-	1 (3.33)

Small and medium families were found to be predominant in all the three selected groups with 40 per cent of members belonging to each category. However, six members were having large family.

Agriculture was the main occupation of main earner for most of the members' families (57%). About 20 per cent of the members were labourers, while 17 per cent members having small business. Economic status of the members in all three selected SHGs found to be BPL (80%).

Majority of the respondents (57%) were having average annual income less than Rs. 25000 followed by 37 per cent of members having annual income between Rs.25000 and Rs.50000. Family income of 37 and 33 per cent of members found to be between Rs.50000 to Rs.100000 and less than Rs.50000, respectively.

Land holdings refer to the land owned or available with member's family to practice agriculture. Members were classified into six categories on the basis of their size of land holding *viz.* landless, marginal (<1 ha), small (1-2 ha), semi medium (2-4 ha), medium (4-10 ha), large (>10 ha). Most of the members belonged to landless (40%). About 37 and 10 per cent of members belonged to marginal and small categories, respectively. One member's (SHG 2) family landholding found to be large that is quite contrasting considering the homogeneity of the group (SHG 2).

Table 6.3 presents the communication and information sources use pattern of the members of selected three SHGs formed under NABARD's SBLP in Kanker district of Chhattisgarh. It is observed that the personal localite sources (friend, neighbours, fellow farmers, village leaders) are mostly preferred followed by the mass media with overall mean score of 15.23 and 10.90, respectively. The use of personal cosmopolite sources (village level worker, agricultural officer, and other line department personnel) is quite low with overall mean score of 4.93. The use of mass media is more varied as compared to personal localite and personal cosmopolite sources as evident from the higher standard deviation values.

Table 6.3: Communication sources use pattern of the members of selected SHGs under NABARD's SBLP

Communication source	SHG 1 (n=10) Mean score (SD)	SHG 2 (n=10) Mean score (SD)	SHG 3 (n=10) Mean score (SD)	Overall (N=30) Mean score (SD)
Personal localite	17.20 (2.49)	14.50 (3.87)	14.00 (4.64)	15.23 (3.91)
Personal cosmopolite	7.20 (4.52)	4.40 (3.81)	3.20 (5.47)	4.93 (4.80)
Mass media	14.10 (5.65)	10.90 (6.38)	7.70 (6.02)	10.90 (6.39)

Minimum and maximum possible scores in case of each source are 0 and 20, respectively

6.2.2 Socio-Personal, Socio-Economic and Communicational Profile of Members of SHGs under NRLM

The Table 6.4 presents the socio-personal and socio-economic profile of the members of three selected SHGs (SHG 4, SHG 5 and SHG 6) formed under NRLM in Kanker block of Kanker district in Chhattisgarh.

It is observed that most of the group members were young and the remaining were of middle age in case of all three selected SHGs. All the members belonging to each group were female.

All three selected SHGs represented OBC, SC, and ST categories; however, overall presence of both OBC and ST categories were equal i.e. 47 per cent, rest being SC. Both SHG 4 and SHG 5 were dominated with ST while SHG 6 was having more OBC members (90%).

The overall education qualifications of the members of all three selected SHGs were primary (40%) and secondary level (47%). One member in each group was illiterate. Only one member (SHG 4) had education of higher secondary level.

All the three selected group of SHGs were mostly having small and medium families. Most of the member's main earner in her family was found to be working on agriculture and allied sector (70%).

Most of the members in all three selected SHGs were found to be BPL (60% and above). As per the annual income status of the member, 83 per cent of group members were having income below Rs. 50000. Family annual income was found to be within Rs. 100000 for about 70 percent of members.

About 30 per cent of the group members were found to be each in landless, marginal and small land holding categories.

Table 6.4: Socio-personal and socio-economic profile of the members of selected SHGs under NRLM

Characteristics	SHG 4 (n=10) Frequency (%)	SHG 5 (n=10) Frequency (%)	SHG 6 (n=10) Frequency (%)	Overall (N=30) Frequency (%)
1. Age:				
Young (20-40 years)	6 (60)	8 (80)	8 (80)	22 (73.33)
Middle aged (41-60 years)	4 (40)	2 (20)	2 (20)	8 (26.67)
Old (> 60 years)	-	-	-	-
2. Sex:				
Male	-	-	-	-
Female	10 (100)	10 (100)	10 (100)	30 (100)
3. Caste:				
General	-	3 (30)	9 (90)	-
OBC	2 (20)	1 (10)	-	14 (46.67)
SC	1 (10	6 (60)	1 (10)	2 (6.66)
ST	7 (70)	-	-	14 (46.67)
4. Education:				
Illiterate	1 (10)	1 (10)	1 (10)	3 (10.00)
Primary	4 (40)	4 (40)	4 (40)	12 (40.00)
Secondary	4 (40)	5 (50)	5 (50)	14 (46.67)
Higher secondary	1 (10)	-	-	1 (3.33)
5. Family:				
Small (upto 4 members)	5 (50)	4 (40)	3 (30)	12 (40.00)
Medium (5-7 members)	4 (40)	6 (60)	6 (60)	16 (53.34)
Large (>7 members)	1 (10)	-	1 (10)	2 (6.66)
6. Occupation of main earner:				
Agriculture and allied	7 (70)	8 (80)	6 (60)	21 (70.00)
Artisan	-	-	1 (10)	1 (3.33)2
Small business	-	1 (10)	1 (10)	(6.66)
Labourer	-	1 (10)	-	1 (3.33)
Others	3 (30)	-	2 (20)	5 (16.67)
7. Economic status:				
APL	4 (40)	3 (30)	4 (40)	11 (36.67)
BPL	6 (60)	7 (70)	6 (60)	19 (63.33)
8. Annual income of member:				
Upto Rs.25000	7 (70)	4 (40)	1 (10)	12 (40.00)
>Rs.25000 to Rs.50000	2 (20)	4 (40)	7 (70)	13 (43.33)
>Rs.50000 to Rs.75000	-	2 (20)	2 (20)	4 (13.33)
>Rs.75000 to Rs.100000	-	-	-	-
>Rs.100000	1 (10)	-	-	1 (3.34)
9. Annual family income:				
Upto Rs.50000	4(40)	4 (40)	-6 (60)	8 (26.67)
>Rs.50000 to Rs.100000	4 (40)	3 (30)	3 (30)	13 (43.34)
>Rs.100000 to Rs.150000	1 (10)	3 (30)	1 (10)	7 (23.33)
>Rs.150000 to Rs.200000	-	-	-	1 (3.33)
>Rs.200000	1(10)	-	-	1 (3.33)

Contd.

Characteristics	SHG 4 (n=10) Frequency (%)	SHG 5 (n=10) Frequency (%)	SHG 6 (n=10) Frequency (%)	Overall (N=30) Frequency (%)
10. Land holding of family:				
Landless	2 (20)	3 (30)	4 (40)	9 (30.00)
Marginal (<1 ha)	5(50)	2 (20)	2 (20)	9 (30.00)
Small (1-2 ha)	1 (10)	5 (50)	3 (30)	9 (30.00)
Semi-medium (2-4 ha)	-	-	1 (10)	1 (3.33)
Medium (4-10 ha)	1 (10)	-	-	1 (3.33)
Large (>10 ha)	1 (10)	-	-	1 (3.33)

Table 6.5 presents the communication and information sources use pattern of the members of selected three SHGs formed under NRLM in Kanker district of Chhattisgarh. It is observed that the personal localite sources (friend, neighbours, fellow farmers, village leaders) were mostly preferred followed by the mass media with overall mean score of 14.43 and 8.73, respectively. The use of personal cosmopolite sources (village level worker, agricultural officer, and other line department personnel) was quite low with overall mean score of 3.80. The use of mass media was more varied as compared to personal localite and personal cosmopolite sources as evident from the higher standard deviation valucs in most occasions.

Table 6.5: Communication and information sources use pattern of the members of selected SHGs under NRLM

Communication source	SHG 4 (n=10)	SHG 5 (n=10)	SHG 6 (n=10)	Overall (N=30)
	Mean score (SD)	Mean score (SD)	Mean score (SD)	Mean score (SD)
Personal localite	15.50 (2.72)	12.90 (3.60)	14.90 (3.48)	14.43 (3.37)
Personal cosmopolite	5.60 (3.72)	1.80 (3.36)	4.00 (4.52)	3.80 (4.08)
Mass media	11.50 (6.72)	5.70 (4.14)	9.00 (6.58)	8.73 (6.21)

Minimum and maximum possible scores in case of each source are 0 and 20, respectively

6.2.3 Socio-Personal, Socio-Economic and Communicational Profile of Members of SHGs under IWMP

Table 6.6 presents the socio-personal and socio-economic profile of the members of three selected SHGs (SHG 7, SHG 8 and SHG 9) formed under IWMP in Kanker block of Kanker district in Chhattisgarh.

Most of the members both in SHG 7 and SHG 9 were young (20-40 years age), while seven out of ten members of SHG 8 were middle aged (41-60 years). Overall about 67 per cent of the members in the SHGs are young.

All the members of SHG 7 were male while that of SHG 8 and SHG 9 were female.

The members of all three selected SHGs represented OBC, SC and ST categories; however, it was overall dominated by ST category (70%). The representation of all the castes were found to be balanced in case of SHG 9, while the ST dominated both in SHG 8 (90%) and SHG 7 (80%).

Most of the members in all three selected SHGs were having educational qualification up to primary and secondary level. Only one member (SHG 8) is educated up to higher secondary level. One member each in SHG 7 and SHG 9 were illiterate.

Small and medium family was found to be predominant in all three selected SHGs with 43 per cent of members belonging to each category. However, one member each in SHG 7 and SHG 9 were having large family, while two members in SHG 8 were having large family.

Most of the main earners of members' families in case of all three selected SHGs were employed in agriculture and allied sector (73%) followed by the artisan (17%). 80 per cent of the members in all three selected SHGs represent BPL family.

Annual income of most of the SHG members (47%) ranged from Rs.25000 to Rs.50000; while family income of half of the selected SHG member is found between Rs.50000 to Rs.100000.

About 27 per cent of the members were landless, while 63 per cent were marginal and small farmers. Half of the members of SHG 8 were landless.

The study of communication and information sources use pattern of the members of selected three SHGs developed under IWMP in Kanker block of Kanker district in Chhattisgarh (Table 6.7) revealed that the personal localite sources (friend, neighbours, fellow farmers, village leaders) were mostly preferred followed by the mass media. However, use of mass media was perceived more variedly as compared to personal localite and personal cosmopolite sources as evident from the higher standard deviation values. It is interesting to note that the members of SHG 7 being male had used the mass media more as compared to the members of SHG 9 and SHG 8 who were female. The use of personal cosmopolite sources (village level worker, agricultural officer, and other line department personnel) was quite low albeit SHG 8 members had better access.

Table 6.6: Socio-personal and socio-economic profile of the members of selected SHGs under IWMP

Characteristics	SHG 7 (n=10) Frequency (%)	SHG 8 (n=10) Frequency (%)	SHG 9 (n=10) Frequency (%)	Overall (N=30) Frequency (%)
1. Age:				
Young (20-40 years)	9 (90)	3 (30)	8 (80)	20 (66.67)
Middle aged (41-60 years)	1(10)	7 (70)	2 (20)	10 (33.33)
Old (> 60 years)	-	-	-	-
2. Sex:				
Male	10 (100)	-	-	10 (33.33)
Female	-	10 (100)	10 (100)	20 (66.67)
3. Caste:				
General	-	-	-	-
OBC	2 (20)	1 (10)	3 (30)	6 (20.00)
SC	-	-	3 (30)	3 (10.00)
ST	8 (80)	9 (90)	4 (40)	21(70.00)
4. Education:				
Illiterate	1 (10)	-	1 (10)	2 (6.67)
Primary	4 (40)	5 (50)	6 (60)	15 (50.00)
Secondary	5 (50)	4 (40)	3 (30)	12 (40.00)
Higher secondary	-	1 (10)	-	1 (3.33)
5. Family:				
Small (upto 4 members)	3 (30)	5 (50)	5 (50)	13 (43.33)
Medium (5-7 members)	6 (60)	3 (30)	4 (40)	13 (43.33)
Large (> 7 members)	1 (10)	2 (20)	1 (10)	4 (13.34)
6. Occupation of main earner:				
Agriculture and allied	8 (80)	7 (70)	7 (70)	22 (73.33)
Artisan	1 (10)	2 (20)	2 (20)	5 (16.67)
Small business	1 (10)	-	-	1 (3.33)
Labour	-	1 (10)	1 (10)	2 (6.67)
7. Economic status:				
APL	2 (20)	2 (20)	2 (20)	6 (20.00)
BPL	8 (80)	8 (80)	8 (80)	24 (80.00)
8. Annual income of member:				
Upto Rs.25000	-	1 (10)	4 (40)	5 (16.67)
>Rs.25000 to Rs.50000	4 (40)	7 (70)	3 (30)	14 (46.66)
>Rs.50000 to Rs.75000	3 (30)	1 (10)	1 (10)	5 (16.67)
>Rs.75000 to Rs.100000	2 (20)	1 (10)	-	3 (10.00)
>Rs.100000	1 (10)	-	2 (20)	3 (10.00)
9. Annual family income:				
Upto Rs.50000	3 (30)	1 (10)	1 (10)	5 (16.67)
>Rs.50000 to Rs.100000	6 (60)	5 (50)	4 (40)	15 (50.00)
>Rs.100000 to Rs.150000	1 (10)	1 (10)	2 (20)	4 (13.33)
>Rs.150000 to Rs.200000	-	1 (10)	1 (10)	2 (6.67)
>Rs.200000	-	2 (20)	2 (20)	4 (13.33)
10. Land holding of family:				
LandlessMarginal (<1 ha)	-	5 (50)	3 (30)	8 (26.67)
Small (1-2 ha)	6 (60)	1 (10)	2 (20)	9 (30.00)
Semi-medium (2-4 ha)	3 (30)	3 (30)	4 (40)	10 (33.33)
Medium (4-10 ha)	1 (10)	1 (10)	1 (10)	3 (10.00)
Large (>10 ha)	-	-	-	-

Table 6.7: Communication and information sources use pattern of the members of selected SHGs under IWMP

Communication source	SHG 7 (n=10) Mean score (SD)	SHG 8 (n=10) Mean score (SD)	SHG 9 (n=10) Mean score (SD)	Overall (N=30) Mean score (SD)
Personal localite	14.50 (3.06)	13.20 (2.57)	11.30 (2.71)	13.00 (3.01)
Personal cosmopolite	3.00 (3.56)	4.10 (3.25)	2.90 (2.47)	3.30 (3.07)
Mass media	10.40 (7.28)	8.20 (5.18)	4.70 (3.77)	7.77 (5.91)

Minimum and maximum possible scores in case of each source are 0 and 20, respectively

6.2.4 Socio-Personal, Socio-Economic and Communicational Profile of Members of SHGs under ATMA

Table 6.8 presents the socio-personal and socio-economic profile of the members of three selected SHGs (SHG 10, SHG 11 and SHG 12) formed under ATMA in Kanker block of Kanker district in Chhattisgarh.

As evident from the Table 6.8, most of the SHG members were young (77%) and the remaining (23%) were in middle age. All the members of SHG 10 were male, where as that of SHG 11 and SHG 12 was female.

According to classification of caste of the members, ST was dominated (67%) in all three groups. The educational qualifications of members were mostly primary and secondary level (87%).

Members (83%) of all three selected were having of small and medium families. Most of the main earners of members families were employed in agriculture and allied sector (60%) followed by the small business (20%).

The ratio of the APL and BPL family was 30:70. Half of selected members were having less than Rs. 25000 annual income followed by 30 per cent of members having annual income between Rs. 25000-50000. As per the annual family income of SHG member it is found that 40 per cent of members were having up to Rs. 50000- Rs. 100000.

Twenty seven per cent of the group members from the selected SHGs were found to be landless and 40 per cent were marginal followed by 23 per cent small land holding.

Table 6.8: Socio-personal and socio-economic profile of the members of selected SHGs under ATMA

Characteristics	SHG 10 (n=10) Frequency (%)	SHG 11 (n=10) Frequency (%)	SHG 12 (n=10) Frequency (%)	Overall (N=30) Frequency (%)
1. Age:				
Young (20-40 years)	8 (80)	8 (80)	7 (70)	23 (76.67)
Middle aged (41-60 years)	2 (20)	2 (20)	3 (30)	7 (23.33)
Old (> 60 years)	-	-	-	-
2. Sex:				
Male	10 (100)	-	-	10 (33.33)
Female	-	10 (100)	10 (100)	20 (66.67)
3. Caste:				
General	-	-	1 (10)	1 (3.33)
OBC	3 (30)	1 (10)	2 (20)	6 (20.00)
SC	1 (10)	-	2 (20)	3 (10.00)
ST	6 (60)	9 (90)	5 (50)	20 (66.67)
4. Education:				
Illiterate	-	-	1 (10)	1 (3.33)
Primary	4 (40)	3 (30)	4 (40)	11 (36.67)
Secondary	5 (50)	6 (60)	4 (40)	15 (50.00)
Higher secondary	1 (10)	1 (10)	1 (10)	3 (10.00)
5. Family:				
Small (upto 4 members)	4 (40)	3 (30)	2 (20)	9 (30.00)
Medium (5-7 members)	5 (50)	5 (50)	6 (60)	16 (53.33)
Large (8-10 members)	1 (10)	2 (20)	2 (20)	5 (16.67)
6. Occupation of main earner:				
Agriculture and allied	6 (60)	8 (80)	4 (40)	18 (60.00)
Artisan	1 (10)	-	1 (10)	2(6.67)
Small business	1 (10)	2 (20)	3 (30)	6 (20.00)
Labourer	-	-	2 (20)	2 (6.67)
Others	2 (20)	-	-	2 (6.67)
7. Economic status:				
APL	2 (20)	2 (20)	5 (50)	9 (30.00)
BPL	8 (80)	8 (80)	5 (50)	21(70.00)
8. Annual income of member:				
Upto Rs.25000	1 (10)	7 (70)	7 (70)	15 (50.00)
>Rs.25000 to Rs.50000	3 (30)	3 (30)	3 (30)	9 (30.00)
>Rs.50000 to Rs.75000	1 (10)	-	-	1 (3.33)
>Rs.75000 to Rs.100000	3 (30)	-	-	3 (10.00)
>Rs.100000	2 (20)	-	-	2 (6.67)
9. Annual family income:				
Upto Rs.50000	3(30)	5 (50)	4 (40)	12 (40.00)
>Rs.50000 to Rs.100000	3(30)	3 (30)	3 (30)	9 (30.00)
>Rs.100000 to Rs.150000	1(10)	1 (10)	1 (10)	3 (10.00)
>Rs.150000 to Rs.200000	1(10)	-	1 (10)	2 (6.67)
>Rs.200000	2(20)	1 (10)	1 (10)	4 (13.33)
10. Land holding of family:				
Landless	(4-10 ha)	1 (10)	1 (10)	8 (26.67)
Marginal (<1 ha)	2 (20)	1 (10)	5 (50)	12 (40.00)
Small (1-2 ha)	4 (40)	5 (50)	3 (30)	7 (23.33)
Semi-medium (2-4 ha)	2 (20)	3 (30)	2 (20)	1 (3.33)
Medium	1 (10)	-	-	2 (6.67)

The study of communication and information sources use pattern of the members of selected three SHGs developed under ATMA in Kanker block of Kanker district in Chhattisgarh (Table 6.9) revealed that the personal localite sources (friend, neighbours, fellow farmers, village leaders) were mostly preferred followed by the mass media. The use of personal cosmopolite sources (village level worker, agricultural officer, and other line department personnel) was quite low.

Table 6.9: Communication and information sources use pattern of the members of selected SHGs under ATMA

Communication source	SHG 10 (n=10) Mean score (SD)	SHG 11 (n=10) Mean score (SD)	SHG 12 (n=10) Mean score (SD)	Overall (N=30) Mean score (SD)
Personal localite	13.20 (3.01)	15.10 (2.73)	13.40 (3.31)	13.90 (3.04)
Personal cosmopolite	3.70 (2.41)	3.50 (2.37)	1.30 (2.31)	2.83 (2.53)
Mass media	8.90 (5.49)	9.30 (5.77)	9.10 (6.14)	9.10 (5.60)

Minimum and maximum possible scores in case of each source are 0 and 20, respectively

The communicational profile (Fig. 6.1) indicates that personal localite was the most preferred source followed by mass media. It is quite surprising to find low use of personal cosmopolite sources, especially when the SHGs were formed under certain programmes with facilitation of government organisation and NGO officials. Overall information use scenario is found better in SHGs formed under NABARD's SBLP.

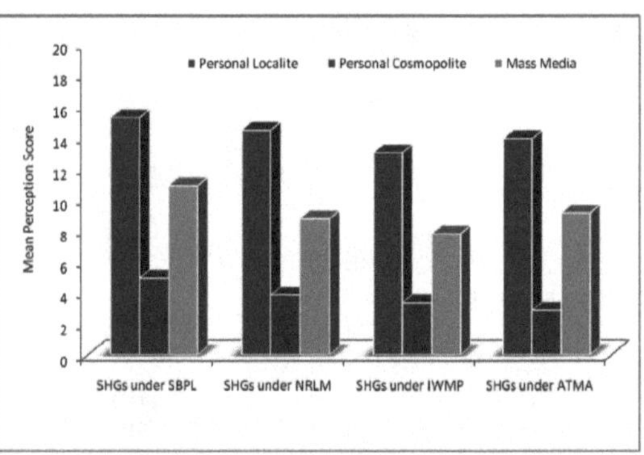

Fig. 6.1: Differential communication sources use pattern of SHG members

The overall socio-personal, socio-economic and communicational profile of the members of SHGs formed under NABARD's SBLP, NRLM, IWMP and ATMA revealed that most of the members (77%) were young aged (20-40 years). Similar findings were reported by Kumaran (1997), Prasad (1998), Puhazhendhi and Jayaraman (1999), Ganesamurthy *et al.* (2000), Sharada (2001), Rao (2005), Gangaiah (2006). However, Banerjee (2002) and Joseph and Easwaran (2006) in their respective studies in Tamil Nadu and Mizoram, respectively, reported that most of the members in the age group of above 40 years participated actively in the group activities.

In the present study, most of the members belonged to ST category (53%) followed by OBC (24%) and SC (20%). The educational level mostly restricted to either primary (41%) or secondary level (42%) with 12% per cent of members was illiterate. Similar findings reported by Ganesamurthy *et al.* (2000), Samar and Raman (2001), Joseph and Easwaran (2006) and Meena and Singh (2012). However, Hemlatha (1995), Kumaran (1997), Suriakanthi (2000), Prita (2001), Savitha (2004) and Rao (2005) observed majority of SHG members were illiterate. Raveendran (2002) reported that 63.3 per cent of group leaders in Tamil Nadu were illiterate whereas in Kerala 58.3 percent got high school education. Sharma *et al.* (2008) stated that raised literacy level could be helpful for the SHGs members to overcome cognition and to understand and gain required skill.

Small and medium family type of most of the members (86%) was evident having agriculture and allied as occupation (65%) of the main earner of family. Prita (2001) and Savitha (2004) mentioned that nuclear families were dominated. Meena and Singh (2012) reported most of SHG family took agriculture as prime occupation (61%). However, several researchers (Puhazhendhi and Jayaraman, 1999; Dadhich, 2001; Dwaraknath, 2001; Puyalavannan, 2001; Samar and Raman, 2001; Balakrishnan, 2002; Jha, 2004; Tripathy, 2004; Bharathi, 2005; Kamaraju, 2005; Gangaiah *et al.*, 2006) mentioned about diversified income generating activities being taken by SHG families, most of which were traditional in nature, non-land/ non-farm based activities and performed at household level. In present study, about 12 and 8 per cent of SHG family members were involved in small business and rural artisanship, respectively.

The members largely belonged to BPL category (73%) with annual income of less than Rs. 50000 (79%) and annual family income less than one lakh (62%) as evident from present study. Joseph and Easwaran (2006) reported similar income status in Mizoram. Meena and Singh (2012) reported that most of the SHG members in Eastern India were having low income.

According to the present study, about one-third of the members (31%) belonged to landless and 34 and 24 per cent had marginal and small land holding. Therefore, the importance of SHG to provide non-land based vocations to the members for income generating activities is quite evident. The diversified activities of the selected SHGs under SBLP of NABARD, NRLM, IWMP and ATMA have already been reported in present study, which hold significance in the context of socio-personal, socio-economic and communicational profile of the SHG members.

6.3 Dynamics and Effectiveness of SHGs

The dynamics of selected SHGs formed under SBLP of NABARD, NRLM, IWMP and ATMA was studied based on the perceptions of sampled SHG members on 10 indicators of group dynamics, while the effectiveness of SHG was assessed based on satisfaction of the SHG members having membership in SHG.

6.3.1 Dynamics and Effectiveness of SHGs under NABARD's SBLP

Table 6.10 presents dynamics of the SHG as perceived by members of selected three SHGs *viz.* SHG 1, SHG 2 and SHG 3, all of which were formed in 2011 under NABARD's SBLP in Kanker block of Kanker district, Chhattisgarh.

The overall dynamics of all three SHGs is varied from 71.40 (SHG 2) to 78.00 (SHG 1) with overall mean score 75.43 out of 100, which may be interpreted as good. All ten indicators were perceived favourably by sampled SHG members. In case of SHG 1, all the indicators were perceived highly (mean score > 6.0), participation being the best (9.50), followed by fund generation (9.00), norms (8.70) and operation & management functions (8.60). Eight out of 10 indicators were perceived highly with mean score >6.0 in case of SHG 2; however, both decision making and membership feeling were perceived as medium (5.70). Fund generation, norms and participation were the best perceived indicators in SHG 3 with all the indicators having mean perception score > 6.0 resulting to overall dynamics of SHG quite good (76.90).

Table 6.10: Dynamics of SHG as perceived by members of SHGs under NABARD's SBLP

Indicators of SHG dynamics	SHG 1 (n=10)		SHG 2 (n=10)		SHG 3 (n=10)		Overall (N=30)	
	Mean score	SD	Mean score	SD	Mean score	SD	Mean score	SD
Participation	9.50	0.85	8.40	0.97	8.70	1.25	8.87	1.11
Decision making	7.00	2.31	5.70	0.95	6.60	1.43	6.43	1.70
Operation & management functions	8.60	0.84	7.40	1.43	7.70	1.16	7.90	1.24
Fund generation	9.00	1.56	9.10	1.10	9.50	0.53	9.20	1.13
Group atmosphere	7.20	0.63	6.60	1.17	7.50	1.58	7.10	1.21
Membership feeling	6.40	1.26	5.70	1.57	6.10	1.45	6.07	1.41
Norms	8.70	1.34	7.80	1.81	9.00	1.49	8.50	1.59
Empathy	6.60	1.51	7.00	1.63	6.80	1.62	6.80	1.54
Interpersonal trust	7.90	1.20	6.20	2.20	7.60	1.84	7.23	1.89
Social support	7.10	1.20	7.50	0.85	7.40	0.84	7.33	0.96
Overall	78.00	5.29	71.40	7.99	76.90	7.23	75.43	7.30

Minimum and maximum possible scores of each indicator are 0 and 10, respectively

Effectiveness of SHG was assessed based on satisfaction of the SHG members having membership in SHG (Table 6.11). The extent of satisfaction was expressed through perceptions of sampled SHG members on five issues *viz.* financial assistance, capacity building, empowerment, living condition and socio-economic status, all of which were perceived very highly, capacity building being the most satisfied issue.

Table 6.11: Satisfaction of SHG members having membership in SHGs under NABARD's SBLP

Issue	SHG 1 (n=10)		SHG 2 (n=10)		SHG 3 (n=10)		Overall (N=30)	
	Mean score	SD	Mean score	SD	Mean score	SD	Mean score	SD
Financial assistance	1.80	0.42	1.80	0.42	2.00	-	1.87	0.35
Capacity building	1.90	0.32	2.00	-	2.00	-	1.97	0.18
Empowerment	2.00	-	1.80	0.42	2.00	-	1.93	0.25
Living condition	1.90	0.32	1.90	0.32	2.00	-	1.93	0.25
Social status	2.00	-	1.80	0.42	2.00	-	1.93	0.25
Overall	9.60	0.97	9.30	1.16	10.00	-	9.63	0.89

Minimum and maximum possible scores of each indicator are 0 and 10, respectively

6.3.2 Dynamics and Effectiveness of SHGs under NRLM

The dynamics of the SHG as perceived by members of selected three SHGs *viz.* SHG 4, SHG 5 and SHG 6, which were formed in 2012, 2007 and 2009, respectively, under NRLM in Kanker block is presented in Table 6.12.

The overall dynamics of SHGs was found high varying from 77.20 in SHG 6, 74.10 in SHG 4 and 71.40 in SHG 5 with overall mean score 75.43 out of 100. All ten indicators were perceived favourably by sampled SHG members. In case of SHG 4, all the indicators were perceived highly (mean score > 6.0), fund generation being the best (8.80), followed by interpersonal trust (8.30) and social support (8.20). Decision making was perceived as medium (5.40) in case of SHG 5, while rest nine 10 indicators were perceived highly, fund generation being the best (9.00) followed by norms (7.70). In SHG 6, fund generation (8.90), group atmosphere (8.30), empathy (8.20) and membership feeling (8.10) were the best perceived indicators resulting to overall dynamics of SHG 6 highest (77.20).

Table 6.12: Dynamics of SHG as perceived by members of SHGs under NRLM

Indicators of SHG dynamics	SHG 4 (n=10)		SHG 5 (n=10)		SHG 6 (n=10)		Overall (N=30)	
	Mean score	SD	Mean score	SD	Mean score	SD	Mean score	S D
Participation	7.80	1.99	6.90	0.99	6.90	0.88	7.20	1.40
Decision making	7.10	1.10	5.40	1.43	7.30	1.06	6.60	1.45
Operation & management functions	7.20	1.14	7.20	1.62	7.20	1.62	7.20	1.42
Fund generation	8.80	1.23	9.00	0.94	8.90	0.74	8.90	0.96
Group atmosphere	6.80	1.14	6.90	1.66	8.30	0.95	7.33	1.42
Membership feeling	6.40	1.65	7.10	1.91	8.10	1.52	7.20	1.79
Norms	7.50	2.84	7.70	1.25	7.60	1.51	7.60	1.92
Empathy	6.00	2.62	7.10	1.66	8.20	1.23	7.10	2.07
Interpersonal trust	8.30	1.06	7.00	1.83	7.80	1.62	7.70	1.58
Social support	8.20	1.03	7.10	0.99	6.90	1.52	7.40	1.30
Overall	74.10	8.44	71.40	5.23	77.20	6.21	74.23	6.96

Minimum and maximum possible scores of each indicator are 0 and 10, respectively

Table 6.13 indicates the effectiveness of SHG that was assessed based on satisfaction of the SHG members having membership in SHG. The extent of satisfaction was expressed through perceptions of sampled SHG members on five issues, all of which perceived quite high, overall empowerment being the most satisfied issue of the SHG members.

Table 6.13: Satisfaction of SHG members having membership in SHGs under NRLM

Issue	SHG 4 (n=10)		SHG 5 (n=10)		SHG 6 (n=10)		Overall (N=30)	
	Mean score	SD	Mean score	SD	Mean score	SD	Mean score	SD
Financial assistance	1.80	0.42	1.90	0.32	1.70	0.48	1.80	0.41
Capacity building	1.60	0.70	1.80	0.42	2.00	0.00	1.80	0.48
Empowerment	1.80	0.42	2.00	-	1.90	0.32	1.90	0.31
Living condition	1.80	0.42	1.90	0.32	1.90	0.32	1.87	0.35
Social status	1.90	0.32	2.00	-	1.60	0.52	1.83	0.38
Overall	8.90	1.29	9.60	0.84	9.10	0.74	9.20	1.00

Minimum and maximum possible scores of each indicator are 0 and 10, respectively

6.3.3 Dynamics and Effectiveness of SHGs under IWMP

Table 6.14 indicates the dynamics of SHG as perceived by members of selected three SHGs *viz.* SHG 7, SHG 8 and SHG 9, which were formed in 2009, 2002 and 2006 respectively, under IWMP in Kanker block of Kanker district, Chhattisgarh. SHG 7 was male SHG, while SHG 8 and SHG 9 were female SHG.

The overall dynamics of SHGs was varied albeit perceived high with overall mean perception score 75.00. It is found highest (83.50) in SHG 8 (female SHG and oldest - since 2002); however, it was relatively low (70.10) in case of another female SHG i.e. SHG 9. The male SHG (SHG 7) showed overall dynamics score 71.40, where decision making and interpersonal trust were perceived medium (mean score <6.0), rest eight indicators being at high level (>6.0). The perceptions of members of SHG 8 for all ten indicators were quite high, norms (9.50), fund generation (9.30) and participation (9.20) being perceived most favourably. The SHG 9 seemed to be suffering from social support (mean perception score 5.60), while other indicators were perceived favourably, fund generation being the highest (9.20) followed by participation (7.90) and operation & management functions (7.90).

Table 6.14: Dynamics of SHG as perceived by members of SHGs under IWMP

Indicators of SHG dynamics	SHG 7 (n=10)		SHG 8 (n=10)		SHG 9 (n=10)		Overall (N=30)	
	Mean score	SD	Mean score	SD	Mean score	SD	Mean score	SD
Participation	8.00	1.25	9.20	0.79	7.90	1.66	8.37	1.38
Decision making	5.30	0.95	7.00	1.41	6.00	0.67	6.10	1.24
Operation & management functions	8.70	0.82	8.20	1.03	7.90	1.10	8.27	1.01
Fund generation	8.80	1.14	9.30	0.48	9.20	0.79	9.10	0.84
Group atmosphere	7.00	1.33	7.70	0.82	6.70	1.25	7.13	1.20
Membership feeling	6.30	1.77	7.90	0.88	6.60	1.65	6.93	1.60
Norms	7.70	0.95	9.50	0.53	6.70	1.83	7.97	1.67
Empathy	6.80	1.32	8.10	1.29	7.50	2.01	7.47	1.61
Interpersonal trust	5.90	1.52	8.40	0.70	6.00	1.25	6.77	1.65
Social support	6.90	1.20	8.20	1.14	5.60	0.97	6.90	1.52
Overall	71.40	4.60	83.50	5.91	70.10	3.78	75.00	7.71

Minimum and maximum possible scores of each indicator are 0 and 10, respectively

Effectiveness of SHG was assessed based on satisfaction of the SHG members having membership in SHG (Table 6.15). The extent of satisfaction was expressed through perceptions of sampled SHG members on five issues, all of which perceived quite high. The financial assistance, empowerment, and socio-economic status were the most satisfying issues for the SHG members.

Table 6.15 Satisfaction of SHG members having membership in SHGs under IWMP

	SHG 7 (n=10)		SHG 8 (n=10)		SHG 9 (n=10)		Overall 8 (N=30)	
	Mean score	SD	Mean score	SD	Mean score	SD	Mean score	SD
Financial assistance	1.90	0.32	2.00	0.00	1.90	0.32	1.93	0.25
Capacity building	1.80	0.42	1.80	0.42	2.00	0.00	1.87	0.35
Empowerment	2.00	-	1.90	0.32	1.90	0.32	1.93	0.25
Living condition	1.90	0.32	1.90	0.32	1.80	0.63	1.87	0.43
Social status	1.90	0.32	2.00	-	1.90	0.32	1.93	0.25
Overall	9.50	0.71	9.60	0.70	9.50	1.27	9.53	0.90

Minimum and maximum possible scores of each indicator are 0 and 10, respectively

6.3.4 Dynamics and Effectiveness of SHGs under ATMA

The dynamics of the SHG as perceived by members of selected three SHGs *viz.* SHG 10, SHG 11 and SHG 12, which were formed in 2011, 2011 and 2012, respectively, under ATMA in Kanker block is presented in Table 6.16. All the members of SHG 10 were male; where as members of both SHG 11 and SHG 12 were female.

Table 6.16 Dynamics of SHG as perceived by members of SHGs under ATMA

Indicators of SHG dynamics	SHG 10 (n=10)		SHG 11 (n=10)		SHG 12 (n=10)		Overall (N=30)	
	Mean score	SD	Mean score	SD	Mean score	SD	Mean score	SD
Participation	5.90	1.20	8.10	1.66	9.10	1.29	7.70	1.91
Decision making	6.50	1.72	7.10	0.88	6.50	1.27	6.70	1.32
Operation & management functions	7.60	0.52	7.50	0.97	7.50	1.35	7.53	0.97
Fund generation	7.10	1.20	8.20	1.03	9.00	1.25	8.10	1.37
Group atmosphere	6.30	0.67	7.50	0.97	6.90	1.37	6.90	1.12
Membership feeling	6.10	1.29	6.20	1.32	5.60	2.07	5.97	1.56
Norms	8.10	1.66	8.40	1.78	7.80	1.87	8.10	1.73
Empathy	8.30	1.06	6.40	1.17	7.10	1.20	7.27	1.36
Interpersonal trust	6.80	1.32	8.20	1.32	6.70	2.26	7.23	1.77
Social support	7.70	0.95	7.00	0.67	7.50	0.85	7.40	0.86
Overall	70.40	2.41	74.60	3.20	73.70	8.10	72.90	5.36

Minimum and maximum possible scores of each indicator are 0 and 10, respectively

Although all the three SHGs showed high level of dynamics, mean perception score varied *viz.* 70.40 in case of SHG 10, 74.60 in SHG 11, 73.70 in SHG 12 indicating that the women SHGs were marginally better than the men SHG formed under ATMA. The members of SHG 10 perceived medium level of participation (5.90), while rest nine indicators were perceived highly (>6.0). In contrast, participation (8.10) was perceived highly by the members of both SHG 11 and SHG 12; other indicators perceived relatively highly were fund generation and norms. However, the members of SHG 12 perceived medium level (5.60) of membership feeling within their group.

Effectiveness of SHG was assessed based on satisfaction of the SHG members having membership in SHG (Table 6.17). The extent of satisfaction was expressed through perceptions of sampled SHG members on five issues, all of which perceived quite high. The socio-economic status was perceived as most satisfying to the SHG members.

Table 6.17: Satisfaction of SHG members having membership in SHGs under ATMA

Issue	SHG 10 (n=10)		SHG 11 (n=10)		SHG 12 (n=10)		Overall (N=30)	
	Mean score	SD	Mean score	SD	Mean score	SD	Mean score	SD
Financial assistance	2.00	-	1.70	0.67	1.70	0.48	1.80	0.48
Capacity building	2.00	-	1.60	0.70	1.90	0.32	1.83	0.46
Empowerment	2.00	-	2.00	-	1.60	0.84	1.87	0.51
Living condition	2.00	-	1.70	0.48	1.90	0.32	1.87	0.35
Social status	2.00	-	2.00	-	1.70	0.67	1.90	0.40
Overall	10.00	-	9.00	1.33	8.80	1.99	9.27	1.44

Minimum and maximum possible scores of each indicator are 0 and 10, respectively

Overall, the dynamics of SHGs was not much varied (overall score ranged from 72.90 to 75.43). SHGs under ATMA showed relatively lesser values for most of the indicators resulting relatively low dynamics as compared to the SHGs under NABARD's SBLP, IWMP and ATMA (Fig. 6.2).The overall dynamics of SHGs studied based on 10 different indicators showed that fund generation, participation and norms were three most effective indicators while decision making, membership feeling and group

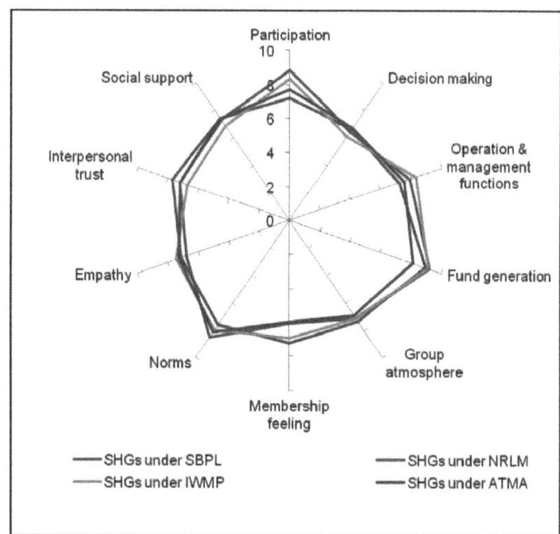

Fig. 6.2: Comparison of dynamics of SHGs under NABARD's SBLP, NRLM, IWMP and ATMA on different indicators

atmosphere were relatively less effective. Vipinkumar and Singh (2001) and Ghosh *et al.* (2010) also reported similar observations in their respective studies on SHGs in Kerala and water user groups in Odisha, respectively. Higher group dynamics was also reported by Vipinkumar and Singh (2002) in their study at Kerala; while, Garai *et al.* (2013) reported medium level of group dynamics in their study at West Bengal.

t-test was done to test difference between perception of the respective members regarding dynamics of SHGs formed under four different programmes *viz.* NABARD's SBLP, NRLM, IWMP and ATMA (Table 6.18).

Table 6.18 Matrix of t-statistic testing difference between perception of members on dynamics of SHGs formed under different programmes

	t- statistic obtained through t-test			
	SHGs under SBLP	SHGs under NRLM	SHGs under IWMP	SHGs under ATMA
SHGs under SBLP	—	0.517	0.824	0.132
SHGs under NRLM	0.517	—	0.688	0.409
SHGs under IWMP	0.824	0.688	—	0.226
SHGs under ATMA	0.132	0.409	0.226	—

t table value at 0.05 level is 2.045

The calculated values of t were less than the table value at 5 per cent level of significance and hence the null hypothesis hold true. Therefore, there were no significant differences between perceptions of members of SHGs formed under four different programmes with respect to self-help group dynamics.

For the test of independence, also known as the test of homogeneity, 'Chi-square (χ^2) test' was conducted. The chi-squared statistic was found as 62.22 that was less than the table value at the 0.05 critical point; therefore, the null hypothesis was accepted. It interprets that the dynamics of SHGs perceived by the members was independent of their SHGs formed under different programmes *viz.* NABARD's SBLP, NRLM, IWMP and ATMA. Contrastingly, Vipinkumar and Asokan (2014) reported the variations in group dynamics in their study in Kerala and most important dimensions were participation, group atmosphere and achievements.

6.3.5 Association between Dynamics of SHG and Members' Characteristics

Association between dynamics of SHG as dependent variable and independent variables like socio-personal, socio-economic, communicational characteristics of SHG members, and satisfaction of SHG members being part of SHG was assessed through correlation analyses (Table 6.19).

It is evident that age, caste, economic status and family land holding were significantly and negatively correlated with dynamics of SHGs under SBLP, NRLM, IWMP and ATMA as well as overall inclusive of all SHGs, which means group dynamics would be more with members of younger age, ST/SC/ OBC caste, low economic status (BPL) and no/ low land holding. On the other hand, use of personal localite communication sources, use of mass media sources, communication/ information use pattern and satisfaction of SHG member being part of her/ his SHG were significantly and positively correlated with dynamics of SHGs under each category as well as overall, which indicates betterment of these independent variables would result in higher group dynamics.

Sex of SHG member in term of female showed overall significant and positive relationship with group dynamics and also the same was evident in case of SHGs under ATMA but not in SHGs under IWMP. All the SHGs both under SBLP and NRLM were female SHGs, thus correlation analyses were not applicable. Education showed significant and positive correlation with dynamics of all SHGs barring the SHGs under IWMP. Overall, correlation coefficient of education was highly significant. Type of family did not show significant association with group dynamics except for the SHGs under NRLM showing negative and significant relationship. Negative and significant relationship of occupation of main earner in family with group dynamics in case of SHGs under both IWMP and ATMA revealed that lower occupational categories such as labourers, marginal and small farmers as the main earner of SHG member's family made the dynamics of her/ his group higher. Annual income of member as well as her/ his family showed positive and significant relationship with dynamics of SHGs both under SBLP and ATMA. Overall correlation coefficients with respect to occupation of main earner in SHG member's family, annual income of SHG member and annual income of SHG member's family were not significant.

Similar types of association between group dynamics and characteristics of members were reported by Garai *et al.* (2013) and Vipinkumar and Asokan (2014) in their respective studies in West Bengal and Kerala.

As 11 out of 15 independent variables showed overall correlation coefficients significant, those were considered as independent variables and dynamics of SHG as dependent variable in step-wise regression analyses to find out the functional relationship (Table 6.20). The insignificant variables were dropped from the analyses in step-wise manner; starting from the least significant variable. Ultimately, at the end of the analyses, the regression model retained

Table 6.19: Correlation of dynamics of SHG with SHG members' characteristics

SHG members' profile/ characteristics	Correlation Coefficient				
	SHGs under SBLP (n=30)	SHGs under NRLM (n=30)	SHGs under IWMP (n=30)	SHGs under ATMA (n=30)	Overall (N=120)
1. Age	-0.486**	-0.405*	-0.437*	-0.510**	-0.501**
2. Sex	—	—	0.102	0.426*	0.259**
3. Caste	-0.480**	-0.466**	-0.383*	-0.487**	-0.368**
4. Education	0.651**	0.596**	0.022	0.564**	0.475**
5. Economic status	-0.698**	-0.696**	-0.388*	-0.484**	-0.483**
6. Type of family	0.135	-0.408*	-0.019	-0.003	-0.049
7. Occupation of main earner in member's family	0.213	-0.076	-0.417*	-0.372*	-0.092
8. Annual income of member	0.412*	0.058	-0.063	0.675**	0.076
9. Annual family income	0.360*	0.026	0.040	0.398*	0.007
10. Family land holding	-0.422*	-0.389*	-0.421*	-0.541**	-0.264**
11. Use of personal localite communication source	0.859**	0.785**	0.583**	0.506**	0.690**
12. Use of personal cosmopolite communication source	0.360*	0.163	0.054	0.199	0.218*
13. Use of mass media communication source	0.608**	0.638**	0.475**	0.390*	0.505**
14. Communication/ information use pattern	0.725**	0.669**	0.511**	0.493**	0.597**
15. Satisfaction of member being part of her/ his SHG	0.686**	0.568**	0.516**	0.570**	0.558**

* Significant at 5 per cent level; ** Significant at 1 per cent level; SHGs both under SBLP and NRLM are female SHGs

the independent variables having significant coefficients up to 5 per cent level of significance. It is revealed that eight out of 11 variables (independent variables) pertaining to characteristics of SHG members were retained in the regression model, *viz.* use of personal localite communication source, economic status, age, sex, satisfaction of member being part of her/ his SHG, use of mass media communication source, caste, and family land holding, having significant t values and in order of their importance in explaining variations in dynamics of SHG (dependent variable). These eight variables together explained about 73 per cent variations (R^2= 0.728) in dynamics of SHG. Almost similar relationship was also observed by Garai *et al.* (2013).

Table 6.20: Regression analyses (step-wise) showing relationship between dynamics of SHG (dependent variable) with SHG members' characteristics (independent variables)

Independent variables	b value	Standard error	t value	F value	R^2
Constant	55.987	5.512	10.158	37.19**	0.728
Use of personal localite communication source	0.737	0.137	5.375**		
Economic status	-2.966	0.914	-3.245**		
Age	-0.189	0.056	-3.366**		
Sex	4.204	0.951	4.422**		
Satisfaction of member being part of her/ his SHG	1.040	0.388	2.683**		
Use of mass media communication source	0.204	0.075	2.720**		
Caste	-0.884	0.426	-2.074*		
Family land holding	-0.209	0.103	-2.017*		

* Significant at 5 per cent level; ** Significant at 1 per cent level

6.4 Impact of SHG on Livelihood of Members

The livelihood security refers to food and nutritional security, economic security, habitat security, educational security, social security and health security. Livelihood is the function of physical, social, financial, human and natural assets. Therefore, impact of selected SHGs formed under SBLP of NABARD, NRLM, IWMP and ATMA was studied based on the perceptions of sampled SHG members on above-mentioned six indicators of livelihood security as well as five indicators of livelihood.

6.4.1 Impact of SHGs under NABARD's SBLP on Livelihood of Group Members

Table 6.21 presents impact of SHGs on livelihood security as perceived by members of selected three SHGs *viz.* SHG 1, SHG 2 and SHG 3, all of which formed in 2011 under NABARD's SBLP in Kanker block of Kanker district, Chhattisgarh.

Table 6.21: Change in livelihood security on joining SHG as perceived by members of selected SHGs under NABARD's SBLP

Indicators of livelihood security	SHG 1 (n=10)		SHG 2 (n=10)		SHG 3 (n=10)		Overall (N=30)	
	Mean score	SD	Mean score	SD	Mean score	SD	Mean score	SD
1. Food & nutritional security								
Before joining SHG	3.10	0.74	3.00	0.47	3.30	0.95	3.13	0.73
After joining SHG	3.90	0.57	3.90	0.57	4.00	0.82	3.93	0.64
2. Economic security								
Before joining SHG	3.00	0.82	3.00	0.67	3.10	0.74	3.03	0.72
After joining SHG	3.80	0.79	3.80	0.63	3.80	0.63	3.80	0.66
3. Habitat security								
Before joining SHG	2.70	0.67	2.50	0.53	2.90	0.57	2.70	0.60
After joining SHG	3.20	0.63	3.70	0.48	3.30	0.48	3.40	0.56
4. Educational security								
Before joining SHG	2.80	0.42	2.80	0.63	2.80	0.63	2.80	0.55
After joining SHG	3.30	0.48	3.60	0.52	3.60	0.70	3.50	0.57
5. Social security								
Before joining SHG	2.60	0.52	2.80	0.79	2.80	0.42	2.73	0.58
After joining SHG	3.50	0.53	3.90	0.74	3.60	0.52	3.67	0.61
6. Health security								
Before joining SHG	3.00	0.67	2.70	0.67	2.60	0.52	2.77	0.63
After joining SHG	3.80	0.92	3.90	0.57	3.90	0.57	3.87	0.68
Overall								
Before joining SHG	17.20	1.93	16.80	1.69	17.50	2.07	17.17	1.86
After joining SHG	21.50	1.96	22.80	1.14	22.20	1.03	22.17	1.49

Minimum and maximum possible scores of each indicator are 1 and 5, respectively

Overall livelihood security of the members was improved on joining the SHGs; maximum improvement was perceived for health security followed by social security. The extent of perceived improvement in overall livelihood security of members was maximum in SHG 2 (from 16.80 to 22.80) followed by SHG 3 (from 17.50 to 22.20) and SHG 1 (from 17.20 to 21.50). In case of SHG 1, improvement of social security of members was highest, while in both SHG 2 and SHG 3, health security of the members was most improved indicator. Even though there was improvement of all indicators of livelihood security, overall livelihood security was at 70-76 per cent level.

Further, impact of SHGs on level of livelihood was also realized based on the perceptions of sampled members of SHG 1, SHG 2 and SHG 3 with respect to the gains of five different assets (Table 6.22). Overall, maximum gain in human asset followed by social asset was visible for the selected SHG members. About 30 per cent improvement in level of livelihood of members was visible in all SHGs, maximum in SHG 2 that may also be attributed to maximum livelihood security of SHG 2 members. In case of both SHG 1 and SHG 3 members, maximum gain was in human asset. In case of SHG 2 members, maximum gain was in financial assets (from 2.40 i.e. below average to 3.70 on 5-point continuum scale). Thus, joining to SHG has improved both livelihood security (29%) and level of livelihood (31%) of rural women.

Table 6.22: Change in level of livelihood on joining SHG as perceived by members of selected SHGs under NABARD's SBLP

Indicators of livelihood	SHG 1 (n=10)		SHG 2 (n=10)		SHG 3 (n=10)		Overall (N=30)	
	Mean score	SD	Mean score	SD	Mean score	SD	Mean score	SD
1. Physical assets								
Before joining SHG	2.70	0.82	2.90	0.57	3.20	0.42	2.93	0.64
After joining SHG	3.70	0.82	3.70	0.48	3.90	0.32	3.77	0.57
2. Social assets								
Before joining SHG	2.80	0.63	3.00	0.67	3.20	0.63	3.00	0.64
After joining SHG	3.80	0.63	4.00	0.67	4.20	0.63	4.00	0.64
3. Financial assets								
Before joining SHG	2.90	0.99	2.40	0.52	3.10	0.32	2.80	0.71
After joining SHG	3.60	0.84	3.70	0.48	3.50	0.53	3.60	0.62
4. Human assets								
Before joining SHG	2.70	0.48	2.90	0.57	2.80	0.42	2.80	0.48
After joining SHG	3.80	0.42	4.10	0.57	4.00	0.67	3.97	0.56
5. Natural assets								
Before joining SHG	3.00	0.67	3.10	0.57	3.30	0.67	3.13	0.63
After joining SHG	4.00	0.82	3.90	0.32	3.90	0.57	3.93	0.58
Overall								
Before joining SHG	14.10	2.56	14.30	0.95	15.60	1.35	14.67	1.83
After joining SHG	18.90	1.97	19.40	0.97	19.50	1.18	19.27	1.41

Minimum and maximum possible scores of each indicator are 1 and 5, respectively

6.4.2 Impact of SHGs under NRLM on Livelihood of Group Members

The impact of the SHG on livelihood security and livelihood as perceived by members of selected three SHGs *viz.* SHG 4, SHG 5 and SHG 6 under NRLM in Kanker block is presented in Table 6.23 and Table 6.24, respectively.

Table 6.23: Change in livelihood security on joining SHG as perceived by members of selected SHGs under NRLM

Indicators of livelihood security	SHG 4 (n=10)		SHG 5 (n=10)		SHG 6 (n=10)		Overall (N=30)	
	Mean score	SD	Mean score	SD	Mean score	SD	Mean score	SD
1. Food & nutritional security								
Before joining SHG	3.10	0.88	2.90	0.57	2.90	0.74	2.97	0.72
After joining SHG	4.00	0.47	4.40	0.70	4.10	0.74	4.17	0.65
2. Economic security								
Before joining SHG	2.80	1.03	3.70	0.48	3.40	0.52	3.30	0.79
After joining SHG	3.60	0.52	4.70	0.48	4.40	0.52	4.23	0.68
3. Habitat security								
Before joining SHG	2.90	0.57	3.00	0.94	3.50	0.85	3.13	0.82
After joining SHG	4.00	0.67	4.20	0.42	4.60	0.52	4.27	0.58
4. Educational security								
Before joining SHG	3.00	0.67	3.30	1.25	2.90	0.88	3.07	0.94
After joining SHG	4.10	0.74	4.10	0.74	4.00	0.67	4.07	0.69
5. Social security								
Before joining SHG	2.90	0.88	3.10	0.74	2.90	0.57	2.97	0.72
After joining SHG	3.60	0.70	4.60	0.70	3.70	0.48	3.97	0.76
6. Health security								
Before joining SHG	2.70	0.82	3.60	0.70	3.20	0.63	3.17	0.79
After joining SHG	3.60	0.70	4.80	0.63	4.30	0.67	4.23	0.82
Overall								
Before joining SHG	17.40	3.06	19.60	2.88	18.80	2.57	18.60	2.90
After joining SHG	22.90	1.79	26.80	1.93	25.10	2.13	24.93	2.49

Minimum and maximum possible scores of each indicator are 1 and 5, respectively

The overall livelihood security of the members was improved in case of all the SHGs, maximum in SHG 5 (from 19.60 to 26.80). Overall, improvement in food & nutritional security and habitat security were perceived relatively high by the members of SHGs leading to livelihood security of members at about 83 per cent level with an improvement of about 34% after joining SHG. In case of SHG 4, both habitat security and educational security of members improved maximum. While in SHG 5, food & nutritional security and social security of members improved maximum. In SHG 6, food & nutritional security of members improved maximum. Thus, on joining the SHG, the members perceived an improvement on their livelihood security which varied from 25 to 37 per cent.

The level of livelihood of the members of selected SHGs before and after joining SHG was also studied on the basis of level of five types of assets holding (Table 6.24). It is quite evident that there were gains in all the assets with an overall improvement in livelihood of members (28%). Overall, the maximum gain (32%) was visible in human and social assets. In SHG 4, members perceived maximum gain in human assets with an overall livelihood improvement of about 31 per cent. Both in SHG 5 and SHG 6, social asset gained the most with an overall 26 and 29 per cent improvement in members' livelihood, respectively. Therefore, both livelihood security and livelihood of members of selected SHGs under NRLM improved to the extent of 34 and 28 per cent, respectively.

Table 6.24: Change in level of livelihood on joining SHG as perceived by members of selected SHGs under NRLM

Indicators of livelihood	SHG 4 (n=10)		SHG 5 (n=10)		SHG 6 (n=10)		Overall (N=30)	
	Mean score	SD	Mean score	SD	Mean score	SD	Mean score	SD
1. Physical assets								
Before joining SHG	2.70	0.48	3.60	0.84	3.50	0.53	3.27	0.74
After joining SHG	3.70	0.67	4.20	0.63	4.40	0.84	4.10	0.76
2. Social assets								
Before joining SHG	2.80	0.92	3.40	0.52	3.10	0.57	3.10	0.71
After joining SHG	3.40	0.52	4.50	0.53	4.40	0.84	4.10	0.80
3. Financial assets								
Before joining SHG	3.20	0.79	3.40	0.84	3.20	0.79	3.27	0.78
After joining SHG	3.90	0.74	4.20	0.42	4.10	0.57	4.07	0.58
4. Human assets								
Before joining SHG	2.80	0.42	3.20	0.92	3.30	0.82	3.10	0.76
After joining SHG	3.90	0.32	4.20	0.63	4.20	0.42	4.10	0.48
5. Natural assets								
Before joining SHG	2.80	0.63	3.20	0.79	3.20	0.63	3.07	0.69
After joining SHG	3.60	0.52	4.00	0.47	4.00	0.82	3.87	0.63
Overall								
Before joining SHG	14.30	2.45	16.80	3.16	16.30	1.49	15.80	2.62
After joining SHG	18.50	0.97	21.10	1.66	21.10	1.73	20.23	1.91

Minimum and maximum possible scores of each indicator are 1 and 5, respectively

6.4.3 Impact of SHGs under IWMP on Livelihood of Group Members

Table 6.25 and Table 6.26 indicate the impact of SHG on livelihood security and livelihood as perceived by members of selected three SHGs *viz.* SHG 7, SHG 8 and SHG 9, which were formed in 2009, 2002 and 2006 respectively, under IWMP in Kanker block of Kanker district, Chhattisgarh. SHG 7 was male SHG, while SHG 8 and SHG 9 were female SHG.

Overall improvement of members' livelihood security is evident in case of all three SHGs; maximum being in SHG 8 (from 16.60 to 22.90) followed by SHG 7 and SHG 9, respectively. It is worth mentioning here that the livelihood was relatively more insecure for the women before joining SHG as evident from the perceptions of members of both SHG 8 and SHG 9. Overall, social security was improved the most as compared to other indicators of livelihood security. In SHG 7, the maximum improvement was in social security followed by habitat security of members. Economic security was the most improved after joining SHG as perceived by the women members of SHG 8. Social security and health security both improved to the similar extent in case of SHG 9 members. Overall livelihood security was improved more in case of women members of SHG 8 (37%) and SHG 9 (33%) as compared to men members of SHG 7 (30%).

The level of livelihood of the members of SHGs was also improved on joining SHG that varied from 32 per cent in SHG 7 (male SHG), 28 per cent both in SHG 8 and SHG 9 (both female SHG). Overall, human asset gain of the members was highest on joining SHG (Table 6.26). It is evident from Table 6.26 that in case of SHG 7, human and financial assets gain of members was more as compared to other assets; while human asset gain was highest in SHG 8 members. Physical asset gain was most in case of members of SHG 9. It is interesting to note here that SHG 8 having all female members showed higher livelihood security as compared to SHG 7 consisting of male members; in contrast, the gain in level of livelihood was more in members of SHG 7 as compared to that of SHG 8. Overall scenario of SHGs under IWMP showed that the overall livelihood security and level of livelihood of the members enhanced by 34 per cent and 29 per cent, respectively.

Table 6.25: Change in livelihood security on joining SHG as perceived by members of selected SHGs under IWMP

Indicators of livelihood security	SHG 7 (n=10)		SHG 8 (n=10)		SHG 9 (n=10)		Overall (N=30)	
	Mean score	SD	Mean score	SD	Mean score	SD	Mean score	SD
1. Food & nutritional security								
Before joining SHG	2.9	0.32	2.80	0.42	2.30	0.48	2.67	0.48
After joining SHG	3.7	0.48	3.90	0.32	3.20	0.42	3.60	0.50
2. Economic security								
Before joining SHG	2.9	0.74	2.40	0.52	2.50	0.53	2.60	0.62
After joining SHG	3.6	0.70	3.80	0.42	3.20	0.42	3.53	0.57
3. Habitat security								
Before joining SHG	2.5	0.53	2.80	0.63	2.70	0.48	2.67	0.55

After joining SHG	3.5	0.71	3.90	0.57	3.50	0.53	3.63	0.61
4. Educational security								
Before joining SHG	2.9	0.57	2.80	0.42	2.70	0.48	2.80	0.48
After joining SHG	3.6	0.70	3.50	0.71	3.40	0.52	3.50	0.63
5. Social security								
Before joining SHG	2.8	0.42	2.80	0.63	2.60	0.52	2.73	0.52
After joining SHG	3.9	0.57	4.00	0.67	3.60	0.52	3.83	0.59
6. Health security								
Before joining SHG	3.1	0.32	3.00	0.67	2.60	0.52	2.90	0.55
After joining SHG	4	0.47	3.80	0.79	3.60	0.52	3.80	0.61
Overall								
Before joining SHG	17.1	0.99	16.60	1.43	15.40	1.17	16.37	1.38
After joining SHG	22.3	1.64	22.90	1.37	20.50	1.27	21.90	1.73

Minimum and maximum possible scores of each indicator are 1 and 5, respectively

Table 6.26: Change in level of livelihood on joining SHG as perceived by members of selected SHGs under IWMP

Indicators of livelihood	SHG 7 (n=10)		SHG 8 (n=10)		SHG 9 (n=10)		Overall (N=30)	
	Mean score	SD	Mean score	SD	Mean score	SD	Mean score	SD
1. Physical assets								
Before joining SHG	2.90	0.57	2.70	0.48	2.50	0.53	2.70	0.53
After joining SHG	3.80	0.63	3.60	0.52	3.60	0.52	3.67	0.55
2. Social assets								
Before joining SHG	3.20	0.79	2.70	0.67	2.80	0.42	2.90	0.66
After joining SHG	3.90	0.74	3.50	0.71	3.70	0.48	3.70	0.65
3. Financial assets								
Before joining SHG	2.40	0.52	2.90	0.32	3.00	0.67	2.77	0.57
After joining SHG	3.50	0.71	3.50	0.53	3.60	0.70	3.53	0.63
4. Human assets								
Before joining SHG	2.90	0.57	2.90	0.32	2.70	0.67	2.83	0.53
After joining SHG	4.00	0.47	4.00	-	3.50	0.53	3.83	0.46
5. Natural assets								
Before joining SHG	3.10	0.74	3.00	0.47	3.10	0.57	3.07	0.58
After joining SHG	3.90	0.57	3.60	0.52	3.60	0.52	3.70	0.53
Overall								
Before joining SHG	14.50	1.27	14.20	1.32	14.10	1.79	14.27	1.44
After joining SHG	19.10	2.02	18.20	1.03	18.00	1.89	18.43	1.72

Minimum and maximum possible scores of each indicator are 1 and 5, respectively

6.4.4 Impact of SHGs under ATMA on Livelihood of Group Members

The impact of the SHG on livelihood security and livelihood of members of selected three SHGs under ATMA in Kanker block is presented in Table 6.27 and Table 6.28. Out of three SHGs *viz.* SHG 10, SHG 11 and SHG 12, which were formed in 2011, 2011 and 2012, respectively, under ATMA in Kanker block, members of SHG 10 were male; where as members of both SHG 11 and SHG 12 were female.

Overall livelihood security of members was improved in all SHGs, maximum (49%) in SHG 10 that was male SHG followed by SHG 11 (22%) and SHG 12 (9%), both of which were female SHGs; however, the level of livelihood security was found relatively higher in case of women members of both SHG 11 and SHG 12 before as well as after their joining to SHG as compared to the male members of SHG 10 (Table 6.27). As ATMA mainly focus on development with respect to the agriculture and allied issues, the SHG of poor male famers felt more secured once they became part of SHG under ATMA. In SHG 10, food & nutritional security of members was improved the most. Habitat security of members was improved in case of SHG 11; while, educational security of members was increased in SHG 12. Overall, with respect to the members of SHGs under ATMA, habitat security improved the most followed by food & nutritional security, economic security, educational security, social security and health security. The extent of improvement in livelihood security of SHG members under ATMA showed relatively less as compared to that of others.

Table 6.27: Change in livelihood security on joining SHG as perceived by members of selected SHGs under ATMA

Indicators of livelihood security	SHG 10 (n=10)		SHG 11 (n=10)		SHG 12 (n=10)		Overall (N=30)	
	Mean score	SD	Mean score	SD	Mean score	SD	Mean score	SD
1. Food & nutritional security								
Before joining SHG	2.20	0.63	3.20	0.63	3.60	1.17	3.00	1.02
After joining SHG	3.80	0.63	3.80	0.42	3.70	0.67	3.77	0.57
2. Economic security								
Before joining SHG	2.00	0.47	3.00	-	3.50	0.53	2.83	0.75
After joining SHG	3.00	0.47	3.80	0.42	3.90	0.74	3.57	0.68
3. Habitat security								
Before joining SHG	1.90	0.32	2.80	0.92	3.50	0.85	2.73	0.98
After joining SHG	2.90	0.32	3.80	0.92	4.10	0.99	3.60	0.93
4. Educational security								
Before joining SHG	2.70	0.48	3.00	0.82	3.20	0.63	2.97	0.67
After joining SHG	3.60	0.52	3.30	0.82	3.80	0.79	3.57	0.73

Contd.

5. Social security								
Before joining SHG	2.30	0.48	3.00	0.67	3.30	0.48	2.87	0.68
After joining SHG	3.30	0.48	3.70	0.82	3.40	1.03	3.40	0.81
6. Health security								
Before joining SHG	2.50	0.71	3.10	0.57	3.50	0.71	3.03	0.76
After joining SHG	3.60	0.84	3.60	0.70	3.60	0.97	3.57	0.82
Overall								
Before joining SHG	13.60	2.17	18.10	1.37	20.60	1.71	17.43	3.41
After joining SHG	20.20	2.25	22.00	1.33	22.50	2.15	21.47	2.10

Minimum and maximum possible scores of each indicator are 1 and 5, respectively

Table 6.28: Change in level of livelihood on joining SHG as perceived by members of selected SHGs under ATMA

ndicators of livelihood	SHG 10 (n=10)		SHG 11 (n=10)		SHG 12 (n=10)		Overall (N=30)	
	Mean score	SD	Mean score	SD	Mean score	SD	Mean score	SD
1. Physical assets								
Before joining SHG	2.20	0.42	3.40	0.52	3.60	0.84	3.07	0.87
After joining SHG	3.30	0.67	3.70	0.67	3.80	0.79	3.60	0.72
2. Social assets								
Before joining SHG	2.40	0.70	3.10	0.57	3.30	0.82	2.93	0.78
After joining SHG	3.40	0.70	3.90	0.88	4.10	0.57	3.80	0.76
3. Financial assets								
Before joining SHG	2.10	0.57	3.10	0.88	3.30	0.82	2.83	0.91
After joining SHG	3.10	0.57	3.90	0.88	4.00	0.67	3.67	0.80
4. Human assets								
Before joining SHG	2.30	0.48	3.00	0.82	3.10	0.74	2.80	0.76
After joining SHG	3.30	0.48	3.80	1.03	3.50	0.97	3.53	0.86
5. Natural assets								
Before joining SHG	2.80	0.42	2.90	0.74	3.00	0.82	2.90	0.66
After joining SHG	3.80	0.42	3.30	0.95	3.20	0.63	3.43	0.73
Overall								
Before joining SHG	11.80	1.93	15.50	1.90	16.30	1.70	14.53	2.67
After joining SHG	16.90	1.97	18.60	1.90	18.60	1.07	18.03	1.83

Minimum and maximum possible scores of each indicator are 1 and 5, respectively

As evident from Table 6.28, the level of livelihood of the members was improved in case of all three SHGs, maximum being in SHG 10 (43%) followed by SHG 11 (20%) and SHG 12 (14%). In SHG 10, the gain of physical assets by members was highest as perceived by the sampled members of present study. Social, financial and human assets were equally gained by the members of SHG 11. In SHG 12, social asset was increased the most for the members. Thus, on joining the SHG overall livelihood security and level of livelihood were increased by 23 per cent and 24 per cent, respectively.

A comparison of impact on livelihood security of members of SHGs under NABARD's SBLP, NRLM, IWMP and ATMA is depicted in Fig. 6.3. It is evident that the extent of impact was realized comparatively more by members of SHGs under NRLM and their present level of livelihood security was also maximum (mean score of 24.93). However, further room for improvement in livelihood security was still there which would be possible with sustainability of SHGs and efforts.

Comparative scenario of level of livelihood of members before and after joining of SHGs under NABARD's SBLP, NRLM, IWMP and ATMA is depicted in Fig. 6.4. The extent of improvement was highest in case of members under SBLP. However, present level of living was highest for members of SHGs under NRLM (mean score of 20.23).

t-tests were done to test difference between perceptions of the respective members regarding their livelihood security and level of livelihood on joining the SHGs formed under four different programmes *viz.* NABARD's SBLP, NRLM, IWMP and ATMA. The calculated values of t were less than the table value (2.045) at 5 per cent level of significance and thus there were no significant differences between perceptions of members of SHGs formed under four different programmes with respect to their improved livelihood security as well as level of livelihood on joining SHGs.

For the test of independence, also known as the test of homogeneity, 'Chi-square (χ^2) test' was conducted. The chi-squared statistic was found as 24.89 in case of livelihood security and 12.87 in case of level of livelihood, which were less than the table value at the 0.05 critical point; therefore, the null hypothesis was accepted. It interprets that the livelihood security as well as level of livelihood perceived by the members were independent of their SHGs formed under different programmes *viz.* NABARD's SBLP, NRLM, IWMP and ATMA.

It is immaterial for rural poor including women that what programme has been supporting them by forming their SHG. But they were satisfied being part of their SHG and have perceived the dynamics of their SHGs quite highly. Moreover, they realized the improvement in their livelihood security although the extent of improvement it varied across the SHGs. The level of their living or livelihood also improved with the gains of different assets *viz.* physical, social, financial, human and natural assets. Most of the members of SHGs perceived the extent of improvement more in case of human assets followed by social assets.

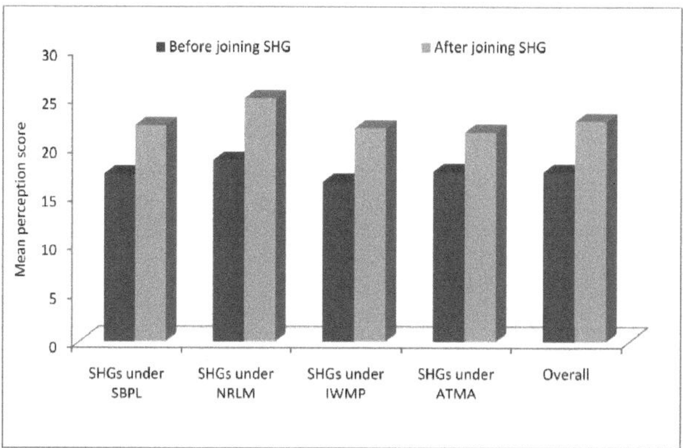

Fig. 6.3: Comparison of impact on livelihood security of members of SHGs under NABARD's SBLP, NRLM, IWMP and ATMA

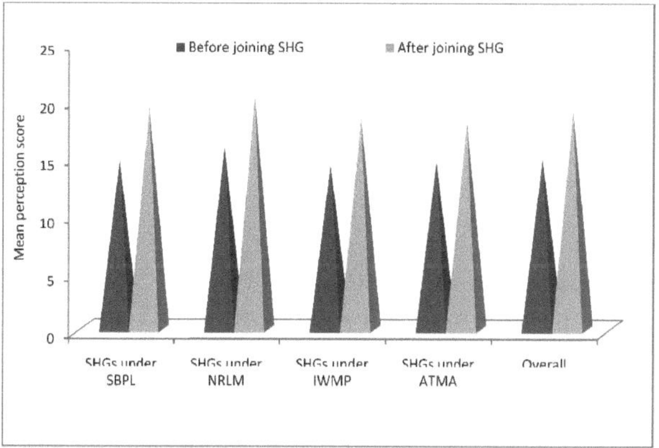

Fig. 6.4: Comparison of impact on level of livelihood of members of SHGs under NABARD's SBLP, NRLM, IWMP and ATMA

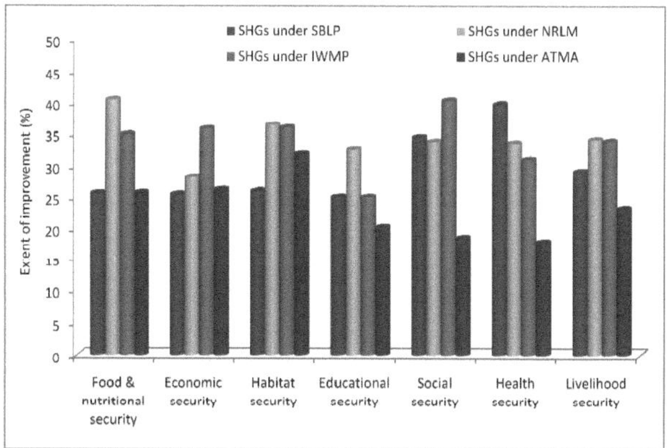

Fig. 6.5: Comparison of improvement in six types of indicators of livelihood security of members of SHGs under NABARD's SBLP, NRLM, IWMP and ATMA

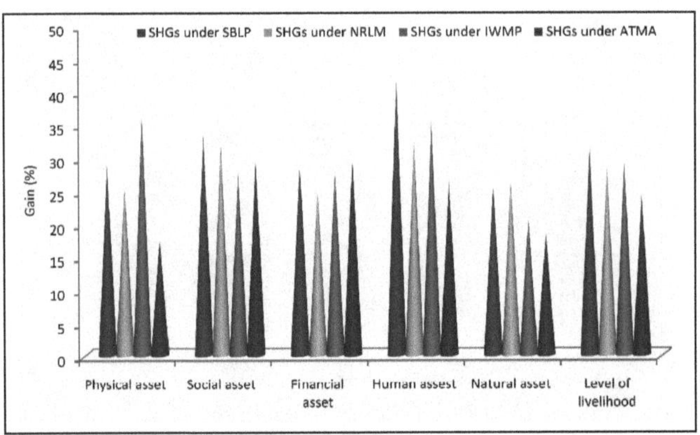

Fig. 6.6: Comparison of gain in five different types of assets by members of SHGs under NABARD's SBLP, NRLM, IWMP and ATMA

Comparison of improvement (%) in six types of indicators of livelihood security of members of SHGs under NABARD's SBLP, NRLM, IWMP and ATMA is depicted in Fig. 6.5. Improvement in food & nutritional security, habitat security and educational security was highest for the members of SHGs under NRLM. Improvement in economic security and social security was highest for the members of SHGs under IWMP. Health security was improved maximum in case of members of SHGs under NABARD's SBLP. Thus, improvement in overall livelihood security was maximum for members of SHGs both under NRLM and IWMP and minimum for members of SHGs under ATMA.

Gain in five different types of assets determining livelihood level of members of SHGs under NABARD's SBLP, NRLM, IWMP and ATMA is depicted in Fig. 6.6. Physical assets gain was highest for the members of SHGs under IWMP. Gain in both social and human assets was highest for the members of SHGs under SBLP. Financial assets gain was maximum for the members of SHGs under ATMA. Natural assets gain was highest for members of both SHGs under SBLP and NRLM. Overall increase in human assets was higher as compared to other types of assets in case of members of SHGs under SBLP, NRLM and IWMP. Thus, improvement in overall level of livelihood was maximum for members of SHGs under SBLP and minimum for members of SHGs under ATMA.

Similar to findings of present study, Srinivasan (1995) illustrated that realization of the poor women that they can take charge of their lives was a significant gain of the SHGs. Puhazhendhi and Jayaraman (1999) reported an improvement in physical, financial and human assets after membership in SHGs. Pitt and Khandker (1998) and Meyer (2001) observed that microfinance through SHG approach contributed to poverty alleviation and food security. NABARD (2002) reported perceptible and wholesome changes in the living standards of SHG members in terms of ownership of assets, borrowing capacities, income generating activities, income levels and increase in savings. Sharma *et al.* (2014) assessed the extent of effectiveness of SHGs in improving livelihood security and gender empowerment that took into accounts both male and female SHGs. Suguna (2006) also remarked that the emergence of SHGs as silent revolution in the spread of rural credit for rural development. Lolheihzovi (2007) considered SHGs as best engine of growth of human resource.

6.5 Influence of SHGs on Empowerment of Rural Poor

Influence of selected SHGs on empowerment of members was studied based on the perceptions of sampled members of selected SHGs under SBLP of NABARD, NRLM, IWMP and ATMA. It was assessed with the help of following indicators *viz.* self development, social empowerment, economic empowerment, political empowerment and information empowerment.

6.5.1 Influence of SHGs under NABARD's SBLP on Empowerment of Group Members

Table 6.29 presents impact of SHGs on empowerment as perceived by members of selected three SHGs *viz.* SHG 1, SHG 2 and SHG 3, all of which formed in 2011 under NABARD's SBLP in Kanker block of Kanker district, Chhattisgarh.

Table 6.29: Influence of SHG on empowerment as perceived by members of selected SHGs under NABARD's SBLP

Indicators of empowerment	SHG 1 (n=10)		SHG 2 (n=10)		SHG 3 (n=10)		Overall (N=30)	
	Mean score	SD	Mean score	SD	Mean score	SD	Mean score	SD
1. Self development								
Before joining SHG	3.10	0.57	2.80	0.63	2.90	0.74	2.93	0.64
After joining SHG	3.70	0.48	3.50	0.53	4.00	0.67	3.73	0.58
2. Social empowerment								
Before joining SHG	2.90	1.10	2.90	0.74	2.80	0.42	2.87	0.78
After joining SHG	3.80	0.63	3.80	0.42	3.80	0.42	3.80	0.48
3. Economic empowerment								
Before joining SHG	2.80	0.42	3.40	0.70	3.00	-	3.07	0.52
After joining SHG	4.00	0.67	4.00	0.47	3.60	0.52	3.87	0.57
4. Political empowerment								
Before joining SHG	3.00	0.82	3.30	0.67	3.10	0.74	3.13	0.73
After joining SHG	3.80	0.79	4.00	0.82	3.90	0.74	3.90	0.76

Contd.

5. Information empowerment								
Before joining SHG	3.45	0.50	3.10	0.66	2.95	0.72	3.17	0.65
After joining SHG	3.85	0.63	3.70	0.48	3.55	0.50	3.70	0.53
Overall								
Before joining SHG	15.25	2.20	12.20	2.20	14.75	1.77	14.07	2.41
After joining SHG	19.15	1.51	15.00	0.67	18.85	1.53	17.67	2.29

Minimum and maximum possible scores of each indicator are 1 and 5, respectively

It is evident that overall empowerment of women members of the SHGs under NABARD"s SBLP was realized. Among the five indicators of empowerment, extent for improvement was maximum in social empowerment. It is also seen previously in the social security as one of the indicators of livelihood security as well as social asset gain as one of the indicators of level of livelihood. About 26 per cent increase in overall empowerment was perceived. SHG 3 created maximum (28%) empowerment followed by SHG 1 (26%) and SHG 2 (23%). However, there is much room to improve the empowerment especially for SHG 2 (overall mean perception score 15 out of 25). In case of SHG 1, members perceived the impact most in term of their economic empowerment followed by social empowerment. In case of SHG 2, social empowerment was realized maximum; while in SHG 3, self development was perceived the most for empowering the women through SHG approach.

Women's empowerment as gender parity as perceived by members of selected SHGs under NABARD's SBLP is depicted in Fig. 6.7. According to this index developed by IFPRI (2012), empowerment as gender parity is measured based on five domains of empowerment with equal weight (20%) and a score of 80 per cent above indicates the women's empowerment from the gender parity point of view. Accordingly, empowerment with gender parity was visible in both SHG 1 and SHG 2. Even though members of SHG 3 perceived highest improvement in empowerment of them but from gender parity point of view it lagged behind with score of 73 per cent. Overall, SHGs under NABARD's SBLP showed empowerment from gender parity (81%). Out of five domains, empowerment was found almost full extent with respect to income and leadership; however, empowerment with respect to time was less.

Differential domains of women's empowerment on the basis of overall perceptions of the members of SHGs under NABARD's SBLP are presented in Fig. 6.8. The sub-domains under each domain get equal weight restricting to overall weight for each domain to 20 per cent so that summing up five domains result to 100 per cent according to this index (IFPRI, 2012). Access to credit under the resource domain was not up to the mark; therefore, it has affected the empowerment with respect to resource from gender parity point of view.

Similarly, lower autonomy in decision making under the production domain resulted lesser empowerment with respect to production domain. Both sub-domains related to time allocation perceived lowly resulting into lesser empowerment of women with respect to time domain.

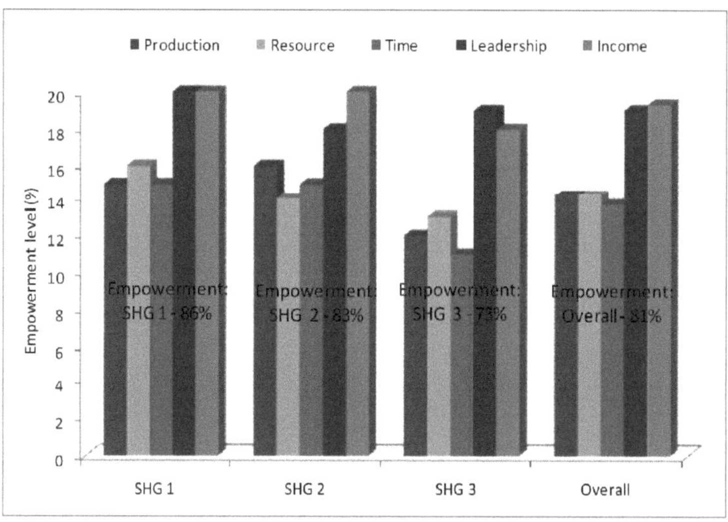

Fig. 6.7: Women's empowerment as gender parity as perceived by members of selected SHGs under NABARD's SBLP

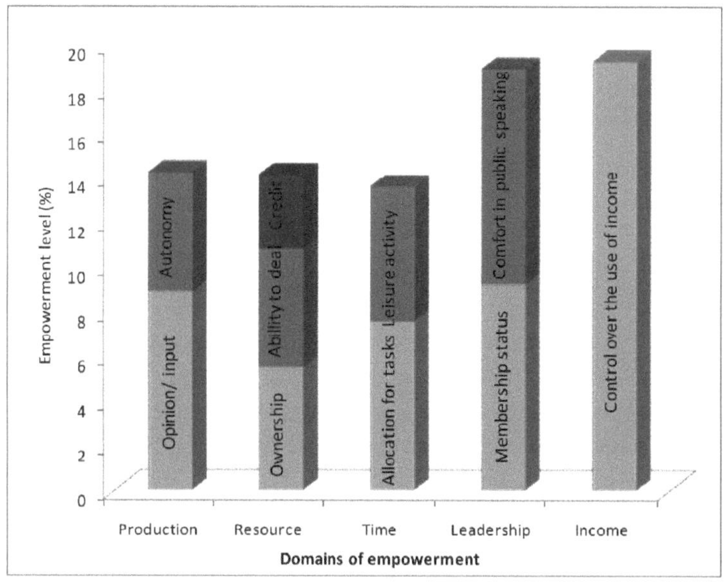

Fig. 6.8: Women's empowerment as gender parity as perceived by members of selected SHGs under NRLM

6.5.2 Influence of SHGs under NRLM on Empowerment of Group Members

The Influence of SHG on Empowerment as perceived by members of selected three SHGs *viz.* SHG 4, SHG 5 and SHG 6, which were formed in 2012, 2007 and 2009, respectively, under NRLM in Kanker block is presented in Table 6.30.

Table 6.30: Influence of SHG on empowerment as perceived by members of selected SHGs under NRLM

Indicators of empowerment	SHG 4 (n=10)		SHG 5 (n=10)		SHG 6 (n=10)		Overall (N=30)	
	Mean score	SD	Mean score	SD	Mean score	SD	Mean score	SD
1. Self development								
Before joining SHG	3.10	0.74	3.00	0.47	2.90	0.57	3.00	0.59
After joining SHG	3.80	0.63	3.90	0.32	4.20	0.42	3.97	0.49
2. Social empowerment								
Before joining SHG	2.60	0.52	3.30	0.82	3.00	0.67	2.97	0.72
After joining SHG	3.70	0.48	4.20	0.42	4.00	0.67	3.97	0.56
3. Economic empowerment								
Before joining SHG	2.60	0.70	3.20	0.42	3.40	0.52	3.07	0.64
After joining SHG	3.70	0.48	4.10	0.32	4.40	0.52	4.07	0.52
4. Political empowerment								
Before joining SHG	2.80	0.92	3.20	0.79	3.90	0.57	3.30	0.88
After joining SHG	3.90	0.57	4.60	0.52	4.70	0.48	4.40	0.62
5. Information empowerment								
Before joining SHG	3.00	0.82	3.05	0.55	3.20	0.54	3.08	0.63
After joining SHG	3.60	0.46	4.25	0.54	4.20	0.42	4.02	0.55
Overall								
Before joining SHG	14.10	2.77	12.55	1.55	16.40	1.35	14.35	2.51
After joining SHG	18.70	1.36	16.45	0.37	21.50	1.08	18.88	2.32

Minimum and maximum possible scores of each indicator are 1 and 5, respectively

It is clear that overall empowerment of women members of the SHGs under NRLM was improved after joining the SHG. Among the five indicators of empowerment, Political empowerment was maximum. It is also to note here that one of the office bearers of SHG 5 selected as *Sarpanch* (Gram Panchayat leadership position) and few others also represent as GP members; thus, the political empowerment as influenced by the SHG was visible. About 31 per cent increase in overall empowerment was perceived. SHG 4 influenced maximum (33%) empowerment followed by SHG 6 and SHG 5 (both 31%). However, there is much room to improve the empowerment especially for SHG 5 (overall mean perception score 16.45 out of 25). The present level of empowerment was found highest in SHG 6 (overall mean score 21.50 out of 30 i.e. 72 %). In case of SHG 4, members perceived the impact most in term of their social, economic and political empowerment. In case of SHG 5, Political empowerment was realized maximum; while in SHG 6, self development of women members was perceived the most influenced by SHG approach.

Women's empowerment as gender parity based on perceptions of members of selected SHGs under NRLM is presented in Fig. 6.9. According to this index of women empowerment developed by IFPRI (2012), empowerment is analyzed on five domains of empowerment in the context of gender parity and a score of 80 per cent and above indicates the women's empowerment. Accordingly, in case of SHGs under NRLM empowerment with gender parity was visible in both SHG 5 and SHG 6 having score of 91 per cent. However, SHG 4 lagged behind with score of 76 per cent. Overall, SHGs under NRLM influenced the empowerment from gender parity (86%). Out of five domains, empowerment was found almost full extent with respect to income and leadership; however, empowerment with respect to time was lowest as per the opinions of sampled women members of selected SHGs under NRLM.

Differential domains of women's empowerment on the basis of overall perceptions of the members of SHGs under NRLM are depicted in Fig. 6.10. The sub-domains under each domain get equal weight restricting to overall weight for each domain to 20 per cent so that summing up five domains result to 100 per cent according to this index (IFPRI, 2012). Both sub-domains related to time domain *viz.* time allocation for productive and domestic task, time for leisure activities were perceived lowly resulting into lesser empowerment of women with respect to time domain. Access to credit and ownership under the resource domain was perceived less; therefore, it has affected the empowerment with respect to resource domain. Similarly, lower autonomy in decision making under the production domain resulted lesser empowerment with respect to production domain. Empowerment in terms of leadership and income domains was almost fullest extent for the women members of SHGs under NRLM.

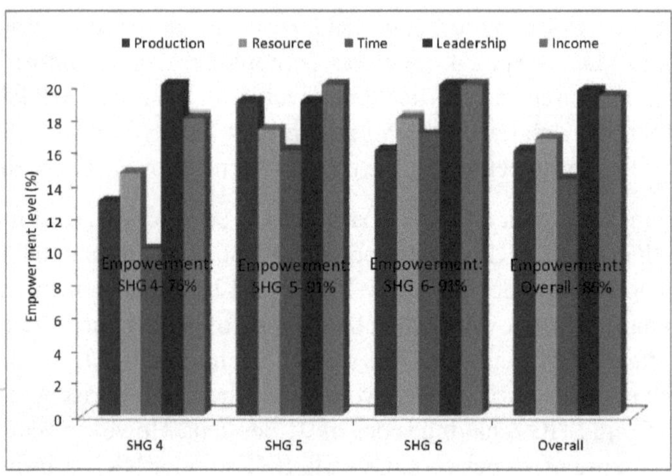

Fig. 6.9: Women's empowerment as gender parity as perceived by members of selected SHGs under IWMP

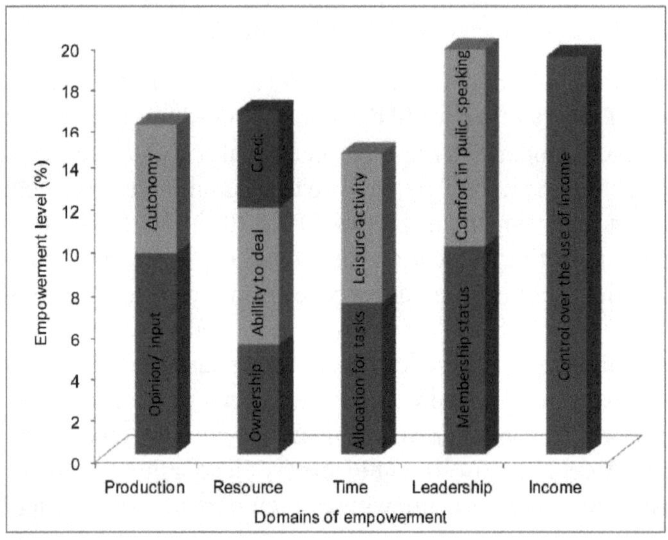

Fig. 6.10: Differential domains of women's empowerment as perceived by members of SHGs under NRLM

6.5.3 Influence of SHGs under IWMP on Empowerment of Group Members

Table 6.31 indicate the influence of SHG on empowerment as perceived by members of selected three SHGs *viz.* SHG 7, SHG 8 and SHG 9, which were formed in 2009, 2002 and 2006 respectively, under IWMP in Kanker block of Kanker district, Chhattisgarh. SHG 7 was male SHG, while SHG 8 and SHG 9 were female SHG.

Table 6.31: Influence of SHG on empowerment as perceived by members of selected SHGs under IWMP

Indicators of empowerment	SHG 7 (n=10)		SHG 8 (n=10)		SHG 9 (n=10)		Overall (N=30)	
	Mean score	SD	Mean score	SD	Mean score	SD	Mean score	SD
1.Self development								
Before joining SHG	3.00	0.82	2.80	0.42	2.40	0.52	2.73	0.64
After joining SHG	3.70	0.67	3.80	0.42	3.40	0.52	3.63	0.56
2.Social empowerment								
Before joining SHG	3.50	0.71	3.10	0.57	2.60	0.52	3.07	0.69
After joining SHG	3.80	0.79	4.10	0.57	3.50	0.53	3.80	0.66
3.Economic empowerment								
Before joining SHG	3.00	0.47	2.70	0.67	2.90	0.57	2.87	0.57
After joining SHG	4.00	0.47	3.50	0.53	3.60	0.52	3.70	0.53
4.Political empowerment								
Before joining SHG	3.40	0.52	2.90	0.57	2.70	0.95	3.00	0.74
After joining SHG	4.10	0.57	4.20	0.42	3.30	0.67	3.87	0.68
5.Information empowerment								
Before joining SHG	3.15	0.53	2.95	0.76	2.75	0.63	2.95	0.65
After joining SHG	3.95	0.37	4.00	0.62	3.40	0.70	3.78	0.63
Overall								
Before joining SHG	16.05	1.69	14.45	1.34	13.35	1.80	14.62	1.93
After joining SHG	19.55	1.12	19.60	1.29	17.20	2.04	18.78	1.87

Minimum and maximum possible scores of each indicator are 1 and 5, respectively

It is evident that overall empowerment of rural poor being members of the SHGs under IWMP was realized through an improvement of 28 per cent. Among the five indicators of empowerment, self development was improved maximum. SHG 8 influenced maximum (36%) empowerment of its members followed by SHG 9 (29%) and SHG 7 (22%). Thus, women members (both SHG 8 and SHG 9) perceived their increased empowerment on joining SHG more as compared to male members (SHG 7). However, there is much room to improve the empowerment further in case of all selected SHGs. In case of SHG 7, male members perceived the impact most in term of their economic empowerment. In case of SHG 8, political empowerment was realized maximum as many of the women members also considered to represent as GP member and other institutions of social and economic importance; while in SHG 9, self development was perceived the most for empowering the women through SHG approach.

Women's empowerment as gender parity as perceived by members of selected SHG 8 and SHG 9 under IWMP is depicted in Fig. 6.11. SHG 7 being the group of male members was not studied. According to this index developed by IFPRI

(2012), empowerment as gender parity is measured based on five domains of women's empowerment with equal weight (20%) and a score of 80 per cent and above indicates the women's empowerment from the gender parity point of view. Accordingly, empowerment with gender parity was visible in both SHG 8 and SHG 9 with score of 80 per cent and 91 per cent, respectively. In case of both these SHGs, members perceived higher improvement in empowerment after joining SHG and they also achieved the empowerment from gender parity point of view. Overall, women SHGs under IWMP showed empowerment from gender parity (86%). Out of five domains, empowerment was found almost full extent with respect to leadership; however, empowerment with respect to time was relatively low.

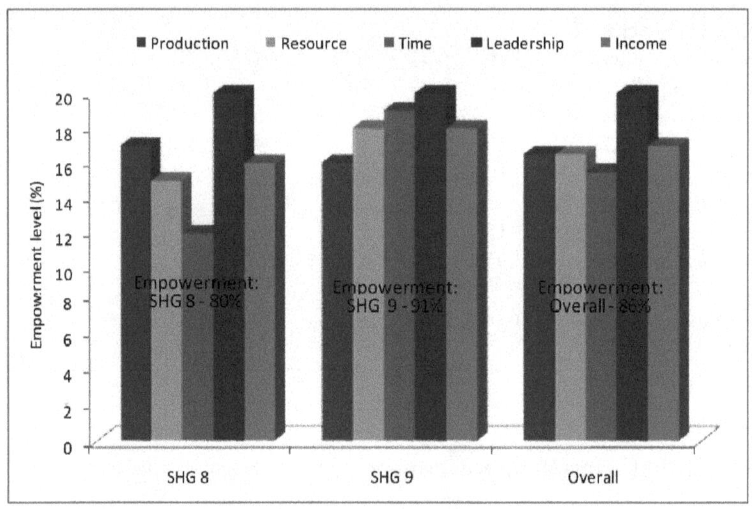

Fig. 6.11: Women's empowerment as gender parity as perceived by members of selected SHGs under ATMA

Differential domains of women's empowerment on the basis of overall perceptions of the members of SHG 8 and SHG 9 under IWMP are presented in Fig. 6.12. According to this index (IFPRI, 2012), the sub-domains under each domain get equal weight restricting to overall weight for each domain to 20 per cent so that summing up five domains result to 100 per cent of empowerment. Both sub-domains related to time allocation to productive and domestic tasks as well as leisure activities perceived lowly resulting into lesser empowerment of women with respect to time domain. Access to credit under the resource domain was perceived low; therefore, it has affected the overall empowerment with respect to resource domain from gender parity point of view. Similarly, lower autonomy in decision making under the production domain resulted lesser empowerment with respect to production domain.

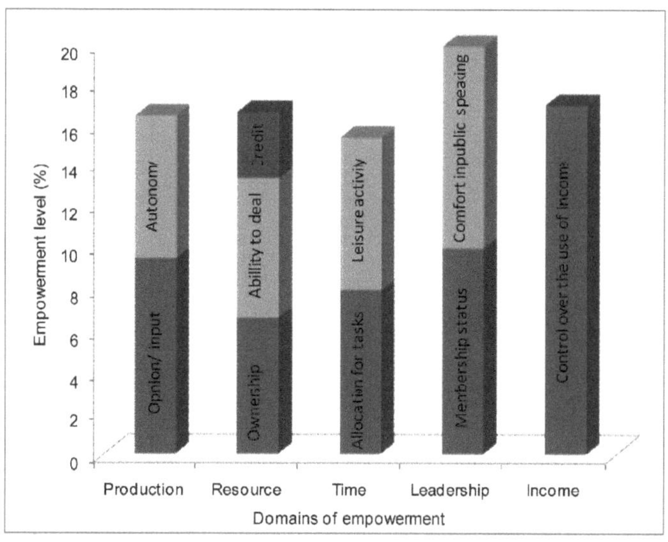

Fig. 6.12: Differential domains of women's empowerment as perceived by members of SHGs under IWMP

6.5.4 Influence of SHGs under ATMA on Empowerment of Group Members

The influence of SHGs on empowerment of members of selected three SHGs under ATMA in Kanker block is presented in Table 6.32. Out of three SHGs *viz.* SHG 10, SHG 11 and SHG 12, which were formed in 2011, 2011 and 2012, respectively, under ATMA in Kanker block, members of SHG 10 were male; where as members of both SHG 11 and SHG 12 were female.

Table 6.32: Influence of SHG on empowerment as perceived by members of selected SHGs under ATMA

Indicators of empowerment	SHG 10 (n=10)		SHG 11 (n=10)		SHG 12 (n=10)		Overall (N=30)	
	Mean score	SD	Mean score	SD	Mean score	SD	Mean score	SD
1. Self development								
Before joining SHG	2.20	0.42	2.80	0.63	3.10	0.99	2.70	0.79
After joining SHG	3.30	0.48	3.80	0.63	3.70	0.48	3.60	0.56
2. Social empowerment								
Before joining SHG	2.90	0.32	2.70	0.48	3.60	0.70	3.07	0.64
After joining SHG	3.90	0.32	3.70	0.67	4.10	0.74	3.90	0.61
3. Economic empowerment								
Before joining SHG	2.10	0.32	3.00	0.67	3.10	0.88	2.73	0.78
After joining SHG	3.10	0.32	3.80	0.63	3.70	0.67	3.53	0.63
4. Political empowerment								
Before joining SHG	2.30	0.48	3.20	0.42	3.30	0.82	2.93	0.74
After joining SHG	3.30	0.48	3.80	0.42	3.70	0.67	3.60	0.56
5. Information empowerment								
Before joining SHG	2.40	0.32	3.30	0.63	3.45	0.76	3.05	0.75
After joining SHG	3.35	0.24	3.55	0.50	3.40	0.70	3.43	0.50
Overall								
Before joining SHG	11.90	1.02	11.80	1.38	16.55	1.61	13.42	2.61
After joining SHG	16.95	1.09	14.85	1.29	18.60	1.78	16.80	2.07

Minimum and maximum possible scores of each indicator are 1 and 5, respectively

It is evident that improvement in overall empowerment of members of the SHGs under ATMA was realized. Among the five indicators of empowerment, self development was maximum followed by social and economic empowerment. About 33 per cent increase in overall empowerment was perceived. It is quite interesting to note that male members of SHG 10 perceived highest improvement (42%) in their empowerment on joining SHG, which may be attributed to the fact that ATMA emphasizes the agricultural activities and rural poor in term of male members felt an opportunity to improve their primary occupation i.e. agriculture being part SHG under ATMA. The improvement in empowerment was 26 per cent in SHG 11 and 12 per cent in SHG 12. There is further scope for improvement in the empowerment of members of selected SHGs under ATMA. In case of SHG 10, male members perceived the impact most in term of their self development. In case of SHG 11, self development as well as social empowerment of women members were realized maximum improvement; while in SHG 12, self development and economic empowerment were perceived more improved by women members as compared to other indicators of empowerment.

Women's empowerment as gender parity as perceived by members of selected SHG 11 and SHG 12 is depicted in Fig. 6.13. According to this index developed by IFPRI (2012), empowerment as gender parity is measured based on five domains of empowerment with equal weight (20%) and a score of 80 per cent above indicates the women's empowerment from the gender parity point of view. SHG 10 being group of all male members was not considered for study. The empowerment with gender parity was not visible in both SHG 11 and SHG 12. Even though members of SHG 11 perceived higher improvement in empowerment of them as compared to members of SHG 12 but from gender parity point of view it lagged behind with score of 79 per cent. Overall, SHGs under ATMA did not achieve empowerment from gender parity (76%). Out of five domains, although empowerment was found high with respect to income and leadership; it was less for other three domains *viz.* production, resource and time.

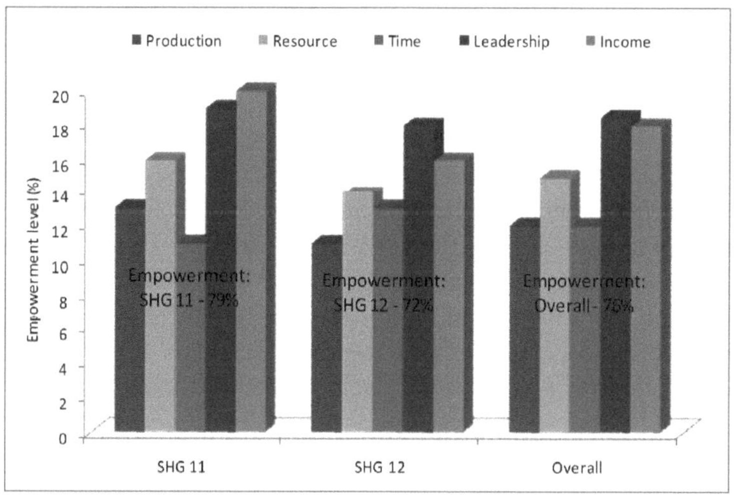

Fig. 6.13: Women's empowerment as gender parity as perceived by members of selected SHGs under ATMA

Differential domains of women's empowerment on the basis of overall perceptions of the members of SHG 11 and SHG 12 under ATMA are presented in Fig. 6.14. The sub-domains under production, resource and time domains showed lower level of empowerment. Access to credit, ownership and ability to deal under the resource domain was not up to the mark; therefore, it has affected the overall empowerment with respect to resource domain. Similarly, lower autonomy in decision making under the production domain resulted lesser empowerment with respect to production domain. Both sub-domains related to time allocation perceived lowly resulting into lesser empowerment of women with respect to time domain.

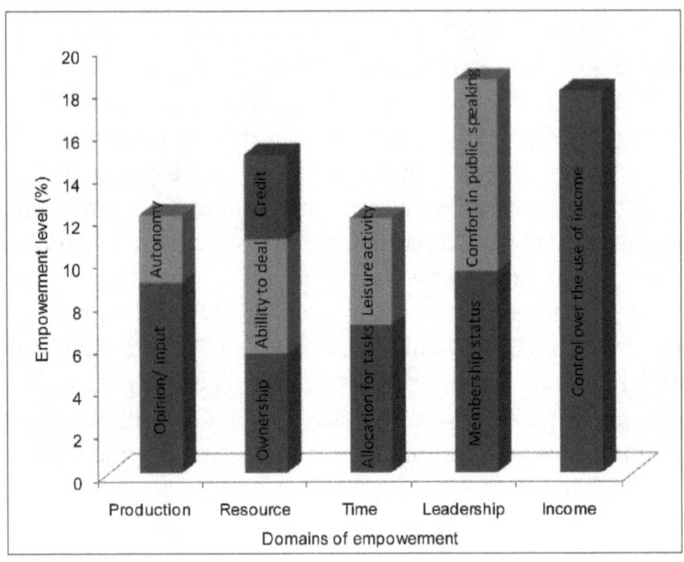

Fig. 6.14: Differential domains of women's empowerment as perceived by members of SHGs under ATMA

t-tests were done to test difference between perceptions of the respective members regarding their empowerment on joining the SHGs formed under four different programmes *viz.* NABARD's SBLP, NRLM, IWMP and ATMA. The calculated values of t (varied from 0.001 to 0.855) were less than the table value (2.045) at 5 per cent level of significance and thus there were no significant differences between perceptions of members of SHGs formed under four different programmes with respect to their improved empowerment.

For the test of independence, also known as the test of homogeneity, 'Chi-square (χ^2) test' was conducted. The chi-squared statistic was found as 18.76, which was less than the table value at the 0.05 critical point; therefore, the null hypothesis was accepted. It interprets that the SHG's influence on empowerment perceived by the members were independent of their SHGs formed under different programmes *viz.* NABARD's SBLP, NRLM, IWMP and ATMA.

A correlation coefficient matrix formulated taking into consideration dynamics of SHG, livelihood security of SHG members, level of living of SHG members and empowerment of SHG members (Table 6.33).

Table 6.33 Correlation coefficient matrix showing association between dynamics of SHG, livelihood security, level of living and empowerment of SHG members

	Dynamics of SHG	Livelihood security	Level of living	Empowerment
Dynamics of SHG	—	0.189*	0.039	0.207*
Livelihood security	0.189*	—	0.549**	0.315**
Level of living	0.039	0.549**	—	0.225*
Empowerment	0.207*	0.315**	0.225*	—

* Significant at 5 per cent level; ** Significant at 1 per cent level; N=120

It is evident that dynamics of SHG, livelihood security of SHG members, and empowerment of SHG members were positively and significantly associated with each other. Dynamics of SHG was not found significantly associated with level of living of SHG members. However, both livelihood security and empowerment were also having significant correlation with level of living.

Similar to the findings of present study, Puhazhendi (2000) in his study of SHGs in the state of Tamil Nadu reported that there is a significant impact of SHGs on social empowerment, empowerment of women, employment generation, credit absorption, new income generating activities, savings pattern *etc*. It is also reported that the creation of income generating assets / activities through loans availed from banks has made significant impact on the overall economic status of the group members. Pandian and Eswaren (2002) reported that participation of the respondents in the decision making process within their families increased after joining SHGs. Puhazhendi and Badatya (2002) pointed out that remarkable improvement was observed in social empowerment of SHG members in terms of self-confidence, better communication and involvement in decision-making, *etc*. Jain (2003) described that SHGs enhanced the equality of status of women as participants, decision makers and beneficiaries in the democratic, economic, social and cultural spheres. Savitha (2004) conducted study on women empowerment decision making in agriculture and reported that the distribution of women according to social empowerment showed that majority had medium social empowerment and 26.67 per cent had high empowerment. NABARD (2005) studied the impact of SHGs on economic empowerment of its member in Ballir district, Uttar Pradesh and reported that there was an increase of the monthly income of each of the families.

Almost similar to the approach of present study on women empowerment from gender parity with the help of an index (IFPRI, 2012), Banerjee and Dutta, (2014) described empowerment as moving from a position of enforced powerlessness to one of the power. There are various indicators that define women empowerment. These indicators are mobility, autonomy, decision making, ownership of household assets, freedom from domination in the family, political and legal awareness, participation in social and development activities, contribution to family expenditure, reproductive rights, exposure to information media and participation in development programmes. The findings of present study find similarity in other studies. Makandar (2011) in a study at Karnataka observed that women have been actively participating in decision making process after becoming members of SHG in the areas consumption of house hold items, education of children and their marriage, number of children, family, planning, purchase and sale of property. NABARD (2011) reported that 76 per cent of the women members were able to interact with officials and 28 per cent of the members were able to save in banks; the result were seen in decision making in household matter, sending children to school, participating in Panchayat election, access to bank credit after joining SHG (98%) as compared to mere two per cent before joining, increase in income by undertaking income generating activities, *etc*. Hussain and Zafar (2012) reported that the financial status of households had improved due to improvement in access across formal credit institutions. Access to credit has enabled women to undertake income generating activities. Sharma *et al.* (2014) assessed the extent of effectiveness of SHGs in gender empowerment and reported that SHGs acted as community platforms from which women become active in village affairs, took part in political decision making process at village level or took action to address social or community issues. Involvement in SHG enabled its members to gain greater control over resources like material possession, knowledge, information, ideas and decision making in home, community, society and nation. Improved awareness of members about social issues was an important indicator of the capacity development of groups. Improved awareness levels enabled group members to play a more effective role in community affairs and worked towards achievement of common goals.

6.6 Overall Impact of SHG Approach

The overall impact of SHG approach was measured considering 10 different issues which are presented in Table 6.34. The issues which were: i) influence over economic resources and participation in economic decision making, ii) influence the individual development and confidence building, iii) influence in decision making of member in the household, iv) improved image of member in society and at home, v) increased capacity building of member, vi) increased awareness and knowledge of member, vii) increased participation and influence of member in social, community and other activities, viii) change in the attitude of the family members/ community regarding SHG member's empowerment, ix) increased mobility, development of networks and interactions with other members of his/ her group and community, x) maintaining economy and equity

Table 6.34: Overall Impact of SHG approach as perceived by members of selected SHGs under different programmes in Kanker district of Chhattisgarh

Impact on member	SHGs under SBLP (n=30)		SHGs under NRLM (n=30)		SHGs under IWMP (n=30)		SHGs under ATMA (n=30)	
	Mean score	SD	Mean score	SD	Mean score	SD	Mean score	SD
1. Influence over economic resources and participation in economic decision making	1.97	0.18	1.93	0.25	2.00	-	1.87	0.43
2. Increased con fidence building	1.93	0.25	1.87	0.35	1.90	0.31	1.87	0.35
3. Influence in decision making in the household 1.93	0.25	1.87	0.43	1.67	0.55	1.80	0.48	
4. Improved image in society and at home	1.87	0.35	1.97	0.18	1.90	0.31	1.77	0.57
5. Increased capacity building	1.93	0.25	1.80	0.55	1.87	0.51	1.80	0.48
6. Increased awareness and knowledge	1.93	0.25	1.97	0.18	2.00	-	1.90	0.31
7. Increased participation and influence in social, community and other activities	1.90	0.31	1.83	0.38	1.93	0.25	1.83	0.53
8. Change in the attitude regarding member's empowerment	1.97	0.18	1.63	0.61	1.83	0.38	1.80	0.48
9. Member's increased mobility, development of networks and interactions	1.93	0.37	1.93	0.25	1.97	0.18	1.90	0.40
10. Maintaining economy and equity in income generating activity among the members	1.97	0.18	1.90	0.40	1.93	0.25	1.77	0.63
Overall	19.33	1.37	18.37	2.27	19.00	1.17	18.30	2.59

Minimum and maximum possible scores of each issue are 0 and 2, respectively in income generating activity among the members. All the issues were very highly perceived by the members of SHGs under NABARD's SBPL, NRLM, IWMP and ATMA; therefore overall impact was very good, mean perception scores ranging from 18.30 in case of SHGs under ATMA to 19.33 in SHGs under NABARD's SBLP.

Manimekalai (2004) remarked that the SHGs have the enough potential for establishing capacity building and self-efficiency among women. Joseph and Easwaran (2006) studied the perceived impact of SHGs on tribal development and found that majority of respondents reported high level of socio-economic impact of SHGs on tribal development. The perceived impact of SHG was found to be significantly associated with three variables duration of membership, members' participation and perceived group cohesion. similar to findings of present study, Dolli (2006) found that the members of SHG had improved income, self employment opportunities, awareness, and social contact Deininger and Liu (2009) reported SHG participation had significant economic impacts.

6.7 Constraints Faced by SHGs

The constraints faced by the selected SHGs formed under different programmes were listed as disclosed by the sampled members of selected SHGs (Table 6.35).

Lack of proper marketing facilities, lack of communication/ information/ facilitation/ cooperation from authorities, inadequate availability of raw material at the right time, unavailability of water during summer hampering group activities like fish farming, *etc*, and lack of monitoring / follow up action by the group promoters were perceived as major constraints by members of SHGs under NABARD's SBLP.

Lack of training and awareness programme, lack of monitoring / follow up action by the group promoters, lack of communication/ information/ facilitation/ cooperation from authorities, membership in more number of groups hampering members effective involvement in group activities, and insufficiency in funding were the major constraints as opined by members of SHGs under NRLM.

Irregular repayment hampering group's thrift and credit activity, unavailability of water during summer hampering group activities like fish farming, *etc*, inadequate availability of raw material at the right time, lack of training and awareness programme, and transportation problems were the perceived major constraints in case of SHGs under IWMP.

Lack of training and awareness programme, irregular/ low production, membership in more number of groups hampering members effective involvement in group activities, insufficiency in funding and lack of communication/ information/ facilitation/ cooperation from authorities were the major constraints as perceived by members of SHGs under ATMA.

Table 6.35: Constraints to SHGs as perceived by the members of SHGs formed under different programmes in Kanker district of Chhattisgarh

Constraint	Rank as perceived by SHG members			
	SHGs under SBLP	SHGs under NRLM	SHGs under IWMP	SHGs under ATMA
1. Inadequate availability of raw material at the right time	III	VII	III	VII
2. Irregular/ low production	VI	VIII	VIII	II
3. Lack of proper marketing facilities of the products	I	IX	IX	VI
4. Unavailability of water during summer hampering group activities like, fish farming, *etc*	IV	XV	II	XV
5. Non-remunerative prices of the products	X	X	X	X
6. Insufficiency in funding	VII	V	XIII	IV
7. Delay in sanctioning/ receiving loan	VIII	VI	XIV	XIV
8. Lack of training and awareness programme	XI	I	IV	I
9. Lack of communication/ information/ facilitation/ cooperation from authorities	II	III	XI	V
10. Membership in more number of groups hampering members effective involvement in group activities	XII	IV	XII	III
11. Irregular repayment hampering group's thrift and credit activity	IX	XI	I	VIII
12. Transportation problems	XIV	XII	V	XI
13. Lack of storage facilities	XIII	XIII	XV	XII
14. Irregular meetings	XV	XIV	VI	XIII
15. Lack of monitoring/follow up action by the group promoters	V	II	VII	IX

VOICE (2008) in a report submitted to Planning Commission mentioned that line department's cooperation was found to be low in majority of the states like Bihar, Uttar Pradesh and Chhattisgarh. The present study also revealed the lack of communication/ information/ facilitation/ cooperation from authorities as one of the major constraints in most of the SHGs. Lack of training and awareness programme was also perceived as one of the major constraints by members of most SHGs. Sharada (2001), Thejaswini *et al.* (2004), and Das (2012) also mentioned this constraint. Inadequate availability of raw material at the right time was found as one of the major constraints in present study. Similar finding was reported by Prita (2001) and Thejaswini *et al.* (2004). Lack of proper marketing facilities, identified in present study, also mentioned as constraint faced by SHGs in the studies conducted by Thejaswini *et al.* (2004), Darlingselvi (2005), Upadhey and Kadam (2008) and Das (2012). Besides, Prita (2001), Thejaswini *et al.* (2004), Joseph and Easwaran (2006), Das (2012) observed many of the constraints similar to the findings of present study.

6.8 Strategy for the Mobilization of Sustainable SHG

Taking into consideration of the valid inferences and practical implications drawn from the findings of the study, a strategy for mobilizing and strengthening an effective and sustainable SHG of rural poor was developed. Steps in this strategy are prioritized through three phases of group development such as group formation phase, stabilization phase and self-helping phase with critical features in each phase:

Group initiation / formation phase (0 to 4 months)

The major steps in this phase should include the initial visit to the location, rapport building, interaction, awareness creation, identification of rural poor including women, introduction meeting, action plan, documentation of deliberations, mobilizing needy members, participatory identification of problems and solutions to problems, action plan for setting group enterprise/ activities including arrangements for required raw materials, marketing information, fortnightly meetings, selection of leader in respect of production, credit and marketing aspects, *etc.*, training on production technology and management aspects emphasizing the maintenance of SHG registers, roles and responsibilities, *etc.*

To facilitate an organization among farmers is in reality a task of great challenge. It is even more so to organize a sustainable functional organisation. In view of that a group promoter (also known as facilitator, community organizer, extension personnel, catalyst or development worker) should prepare himself and community before actually helping in group building. Three major steps and 18 different sub-steps have been identified under group initiation / formation stage.

A. Learning about the village / community / project area

1. Acquiring initial information of the project area
2. Entering the area, discussing with key persons/influential people and introducing self
3. Holding a public meeting for awareness creation about the programme/ project
4. Gathering more information
5. Building people's confidence and community's trust

The most important message for fulfilling this step is 'equipping oneself with as much local knowledge as possible on various aspects, such as political, social, cultural, traditional, religious, ecological, economical and traditional village organizations'. Support of local leaders is to be achieved first. So, discussion is mandatory with the influential people in the community. The local leaders may assist and facilitate to organize a meeting to introduce oneself and the project to all the villagers/farmers. Comments and suggestions from them should be invited in meeting. Some people, especially women, may be shy or afraid of discussing at the meeting. If one finds that women are too reluctant to participate in discussion, a separate meeting for them should be held to facilitate their confidence and convenience. A single meeting is never enough to gather sufficient and reliable information on the community; several meetings, informal as well as formal, need to be organized. Gaining the trust and confidence of farmers is a hard and time-consuming task. But, it plays a pivotal role in building users group. Generally, most farmers are from poor families and always face some troubles and problems. If one pays attention to them, listen to their complaints about hardship and problems with sympathy and try to help them whenever possible, their confidence and trust will grow.

B. Development of local understanding

1. Participatory awareness building of issues related to resources.
2. Conceptualization of local geographical features.
3. Participatory need assessment and prioritization of common needs.
4. Analysis of causative problems and generating ideas of developing solutions and managing resources on a self-help basis.
5. Highlighting advantages of working together in groups and encouraging for formation of group.

6. Motivating the farmers by explaining that in a group, all members of group benefit from their combined skill and resources.
7. Arranging observation tour to witness success stories of other places.

The purpose of this step is to make awareness among farmers of their own community issues related to resources and problems through participatory discussion. From such discussions and debates, they can be benefited by learning about their community's resources and their utilization, problems and needs; their misconceptions can also be eliminated. Farmers will notice how good or bad resources are being utilized and be able to envision appropriate use of resources for areas based on their own local knowledge. The idea is to how to develop and manage the resources on a self-help basis can evolve. Farmers are to be encouraged to tell most important needs with respect to them. Although individual needs may differ, some needs may be common. If the listed common needs are numerous, one cannot deal with all at one time. So, prioritization of common needs will identify the need that requires top most priority. After getting the priority list, one has to start analyzing from the topmost need together with the farmers. It is imperative that the majority reaches a consensus on the selection of the problems that are to be solved urgently so that a commitment is made to remove their problems by them only. Hence, they have to bear in mind that the strengthening or formation of farmers' / users group can tackle their own problems by themselves as per their own interests, all members of group benefit from their combined skill and resources. Success-stories of other places in this regard will also motivate them.

C. Formation of group

After the completion of aforementioned steps, farmers can be reckoned to have been furnished with preparatory measures for formation of their group. The following is a process for group formation:

1. Holding formal meetings to establish group
2. Setting specific objectives of group
3. Determining group's size
4. Structuring group through selection/election of office bearers of the group
5. Establishing group norms/rules
6. Determining functions and work plans though participatory discussion amongst group members.

This step completes the process of group formation that starts with a formal plenary meeting of the farmers in presence of village level government personnel. After the vision and mission of users group are well defined and understood, the group members describe in a clear way specific objectives and expectations in the context of group objectives and activities. There is no hard and fast rule defining the best size of a group since it varies from case to case. It is hard to say how many members should be included in a group. In practice the group size would depend on degree of participation and population of the community. Group structuring with facilitation of the group promoter elect the office bearers as desired by all members. It is also imperative for the group to lay down rules and regulations. The work plan should be prepared though participatory discussion amongst group members. What to do, when to do and who is responsible to do need be stipulated in the work plan so that group can monitor various activities, review progress and find ways and means to overcome problems and difficulties if encountered.

Group stabilization phase (4 to 15 months)

This phase should involve regular need based fortnightly meetings, maintenance of documents, practical trainings of leaders on production, credit and marketing aspects, *etc.*, scheduled implementation of action plan, procurement of inputs, production based on market demand, market synchronized production planning, intensive training to carry out activities of production, credit and marketing aspects, training to other members by the leaders and changing the leadership of SHG after two years so that periodic rotation develops the leadership giving the other potential leaders a chance thereby maintaining the intervening agency's role as 'enabler.'

Most of the cases project personnel are remained active up to the group formation (0-4 months) stage. However, to ensure the stability of the group, group promoter should become 'Enabler' in this stage. Three major steps and 15 sub-steps have been formulated for group stabilization phase.

A. Enabling the group

1. Fulfilling legal aspects to establish the group as statutory body (Registration)
2. Developing habits of team-work and inculcate commitment among group members; expected outcomes should be
   - People willing to go along
   - Ready to share responsibility
   - Plan of action emerges

3. Inculcating a feel for the group dynamics i.e. remaining alert for the various indicators of avoidance is a must among the group members
4. Training in all technical and management aspects inclusive of group's fund generation, accounting and record keeping
5. Capacity building through awareness camps, exposure visits, participatory learning and action.

B. Empowering the group

1. Vesting all decision making power to the group
2. Giving responsibility of operation, maintenance and management of resources to group in its jurisdiction
3. Maintaining group fund for financial independency/self sufficiency
4. Formulation and implementation of action plan taken up by the group
5. Encouraging monitoring and evaluation
   - Leader/office bearer of group monitor and evaluate
   - Members share assessment
   - Success satisfies and failures yield lessons.

C. Developing self-reliance

1. Establishing effective link with services for access
2. Acquiring necessary skills and expertise for efficient operation, maintenance and management of resources
3. Solving problems of own
4. Being keen on next venture
5. Playing proactive role - greater the initiative taken by member-farmer(s) better it is.

Self helping phase (15 to 36 months)

The main steps to be included in this phase are development of fortnightly action programme, meetings for sharing experiences, refinement, improvement and problem solving for the activities under the responsibilities of the leaders, limiting the development personnel's role to a facilitator, gradually reducing their presence at meetings, transactions fully within the SHG under the members themselves, originally active leaders giving way to new leaders after two year term, encouraging inter-SHG contacts, developing a healthy competition spirit, initiative by leaders to create a sense of group pressure by established norms for defaulters, ensuring participation in every activity for sustenance of SHG, ensuring favorable group atmosphere, functions, empathy and inter persona l trust for significant achievements of SHG, emphasizing all the dimensions of group dynamics.

Group promoting personnel gradually reduce their presence during this stage, as group of farmers can stand up on its own, without the help from outside. Two major steps and 12 sub-steps have been conceptualized for this stage.

A. Ensuring independent group / organisation

1. Developing action programme for performance enhancement of the group and group members
2. Selection of master farmers in respect of irrigation management, technical, production, marketing and credit aspects to establish separate unit under each- master farmer's committee
3. Skill development of selected master farmers through training in respective areas
4. Imparting intensive technical knowledge to master farmers to carry out activities of each unit and equipping them to give training to improve knowledge, attitude and skill of other farmers
5. Planning and execution of programmes of action plan as per schedule of different units
6. Organising periodical meetings of group members for farmer to farmer technology transfer – farmer led extension
7. Reducing and facilitating role of group promoters/project personnel for the refinement, improvement and problem solving with respect to various activities through master farmers
8. Development of new set of master farmers after one to one and half year paving the way for leadership development among the group members in due course.

B. Sustainability of group

1. Assuring active participation of members in every activity of group
2. Follow-up actions in post operational/post-project stage

3. Setting appropriate mechanisms to resolve conflicts / problems within the group
4. Establishing networking of farmers' organizations/groups.

Most of the cases project personnel are concerned on the first two stages viz. group formation (0-4 months), group stabilization (4-15 months) and the last stage i.e. group-independency phase (15-36 months) remains largely unaddressed leading to the unsustainability of the group. While helping to build the farmers' groups their prolonged integrity and functionality should be always kept in mind. All the efforts and resources invested in forming groups will be meaningless if they do not sustain themselves for long. Therefore, it is of paramount importance to keep it functional and effective for a long time. In fact it should become a part of the tradition of the village over time, as is the case with the already existing traditional village organizations. Passive participation (for gaining subsidy or other monetary inputs or food for work or due to project pressure *etc.* only) will not keep it sustainable and functional for long. Soon after a project, such groups or organizations become defunct and get dismantled. In other words, only active participation for self-development of farmers and for their own motivation can help to create an effective and sustainable organization. There must be some higher motive for the farmers to participate in a program.

Success and achievement of group depend on the extent to which nature and functioning of the programme/project address the problems and needs of the farmers, the extent to which the farmers have been organized in group with participation and empowerment culture for group action, the extent to which the improvements can be made in the strategies for effective group mobilization and sustainability.

Chapter 7

SHG: Looking Back and Looking Ahead

The main purpose of this chapter is to summarize the results to draw the conclusions on the basis of the foregoing analyses and to indicate some of the significant implications of the present study.

In previous Chapter, the study described a detail account of 12 SHGs, three SHGs from each of four schemes *viz.* NABARD's SHG-Bank Linkage Programme (SBLP), National Rural Livelihood Mission (NRLM), Integrated Watershed Management Programme (IWMP), Agricultural Technology Management Agency (ATMA) in Chhattisgarh state. However, looking back to this study with the salient findings and implications would be the precursor for looking ahead.

7.1 Salient Findings

The major findings of the study are summarized below:

1. The diversified activities were performed by the selected SHGs under NABARD's SBLP, NRLM, IWMP and ATMA, which hold significance in the context of profile of the SHG members, their livelihood and empowerment.
2. Majority of SHG members (77%) were young aged (20-40 years), belonged to ST category (53%) followed by OBC (24%) and SC (20%), having primary (41%) or secondary level (42%) education with 12% per cent of members as illiterate. Small and medium family type of most of the members (86%) was evident having agriculture and allied as occupation (65%) of the main earner of family. The members largely belonged to BPL category (73%) with annual income of less than Rs. 50000 (79%) and annual family income less than one lakh (62%). According to the present study, about one-third of the members (31%) belonged to landless and 34 and 24 per cent had marginal and small land holding.

3. The overall dynamics of SHGs studied based on 10 different indicators showed that fund generation, participation and norms were three most effective indicators while decision making, membership feeling and group atmosphere were relatively less effective. Overall, the dynamics of SHGs was not much varied (overall score ranged from 72.90 to 75.43). SHGs under ATMA showed relatively lesser values for most of the indicators resulting relatively low dynamics as compared to the SHGs under NABARD's SBLP, NRLM and IWMP. There were no significant differences between perceptions of members of SHGs formed under four different programmes with respect to self-help group dynamics as evident from t-tests. Chi-square (χ^2) test of independence also showed the dynamics of SHGs perceived by the members was independent of their SHGs formed under different programmes.

4. Among the characteristics of SHG members, age, caste, economic status and family land holding were significantly and negatively correlated with dynamics of SHGs under SBLP, NRLM, IWMP and ATMA as well as overall inclusive of all SHGs, which means group dynamics would be more with members of younger age, ST/SC/OBC caste, low economic status (BPL) and no/ low land holding. On the other hand, use of personal localite communication sources, use of mass media sources, communication/ information use pattern and satisfaction of SHG member being part of her/ his SHG were significantly and positively correlated with dynamics of SHGs under each category as well as overall, which indicates betterment of these independent variables would result in higher group dynamics. In the step-wise regression analyses, eight variables pertaining to characteristics of SHG members *viz.* use of personal localite communication source, economic status, age, sex, satisfaction of member being part of her/ his SHG, use of mass media communication source, caste, and family land holding, together explained about 73 per cent variations (R^2= 0.728) in dynamics of SHG.

5. Overall livelihood security of members was improved in all SHGs, the extent of improvement was comparatively more for members of SHGs under NRLM (34%) and IWMP (34%) followed by SBLP (29%) and ATMA (23%). Present level of livelihood security was also maximum in case of members of SHGs under NRLM (mean score of 24.93 out of 30) followed by NABARD's SBLP (22.17), IWMP (21.90) and ATMA (21.47). Improvement in food & nutritional security, habitat security and educational security was highest for the members of SHGs under NRLM. Improvement in economic security and social security was highest for the

members of SHGs under IWMP. Health security was improved maximum in case of members of SHGs under NABARD's SBLP. t-tests showed that there were no differences between perceptions of the respective members regarding their livelihood security on joining the SHGs formed under four different programmes. Chi-square (χ^2) test of independence also showed the same result.

6. The level of livelihood of the members was improved in case of all SHGs, extent of improvement was highest in case of members under SBLP (31%) followed by IWMP (29%), NRLM (28%) and ATMA (24%). However, present level of living was highest for members of SHGs under NRLM (mean score of 20.23 out of 25) followed by NABARD's SBLP (19.27), IWMP (18.43) and ATMA (18.03). Physical assets gain was highest for the members of SHGs under IWMP. Gain in both social and human assets was highest for the members of SHGs under SBLP. Financial assets gain was maximum for the members of SHGs under ATMA. Natural assets gain was highest for members of both SHGs under SBLP and NRLM. Overall increase in human assets was higher as compared to other types of assets in case of members of SHGs under SBLP, NRLM and IWMP. t-tests showed that there were no differences between perceptions of the respective members regarding their level of livelihood on joining the SHGs formed under four different programmes. Chi-square (χ^2) test of independence also showed the same result.

7. Improvement in overall empowerment of members of the SHGs was realized, the highest level of empowerment was in case of members of SHGs under NRLM (mean score of 18.88 out of 25), followed by IWMP (18.78), SBLP (17.67) and ATMA (16.80). However, the extent of improvement was maximum in case of members of SHGs under ATMA (33%) followed by NRLM (31%), IWMP (28%) and NABARD's SBLP (26%). Among the five indicators of empowerment, overall social empowerment and political empowerment was improved the most in members of SHGs under SBLP and NRLM, respectively. However, self development was improved the most in members of both IWMP and ATMA. t-tests showed that there were no differences between perceptions of the respective members regarding their empowerment after joining the SHGs formed under four different programmes. Chi-square (χ^2) test of independence showed that SHG's influence on empowerment perceived by the members were independent of their SHGs formed under different programmes.

8. Empowerment from gender parity point of view was studied in case of women SHGs (10 out of 12 SHGs selected for present study) on the basis of five domain of empowerment. Overall, it was found that members of six SHGs (two each under NRLM, IWMP and SBLP) were empowered from gender parity having score of 80 per cent and above. The empowerment from gender parity was highest for the women members of SHGs under both NRLM (86%, range 76-91%) and IWMP (86%, range 80-91%) followed by SBLP (81%, range 73-86%). The members of SHGs under ATMA were not considered empowered from gender parity point of view as the score was less than 80 percent (76%, range 72-79%). Leadership and income domains showed the maximum empowerment in case of women members of all SHGs; however, it was varied and relatively less in case of production, resource, and time domains.
9. Dynamics of SHG, livelihood security of SHG members, and empowerment of SHG members were positively and significantly associated with each other having significant correlation coefficient values. Dynamics of SHG was not found significantly associated with level of living of SHG members. However, both livelihood security and empowerment were also having significant correlation with level of living. Overall impact was perceived very highly by the members of selected SHGs.
10. The major constraints faced by the SHGs varied. Lack of communication/ information/ facilitation/ cooperation from authorities, lack of training and awareness programme, inadequate availability of raw material at the right time, lack of proper marketing facilities of the products, irregular repayment hampering group's thrift and credit activity, lack of monitoring / follow up action by the group promoters, irregular/ low production, *etc* were perceived as some of the major constraints for SHGs. Taking into consideration of the valid inferences and practical implications drawn from the findings of the study, a strategy for mobilizing and strengthening an effective and sustainable SHG of rural poor overcoming the existing constraints was developed.

7.2 Conclusion

It made no difference for rural poor including women that what programme had been supporting them to form their SHG. But they were satisfied being part of their SHGs and have perceived the dynamics of their SHGs quite highly. Moreover, they realized the improvement in their livelihood security after joining the SHGs, although the extent of improvement varied across the SHGs. The level of their living or livelihood also improved with the gains of different assets *viz.* physical, social, financial, human and natural assets on joining the SHGs.

Most of the members of SHGs perceived the extent of improvement more in case of human assets followed by social assets. The empowerment of rural poor including the women was achieved to a great extent in case of all SHGs. However, the women's empowerment from gender parity was achieved in case of six out of ten women SHGs that showed the need of more empowerment of women especially with respect to production, resource and time domains. The significant association of most of the characteristics of SHG member with dynamics of SHG underlined the importance of members to keep SHGs vibrant with effective functioning. The dynamics of SHGs not only depended on profile of the members but also satisfaction of members being part of SHGs and context of formation as well as area of operation of SHGs. The dynamics of SHGs, livelihood security of members and empowerment were three crucial facets of SHG approach, which found to be significantly associated with each other. The constraints faced by the SHGs used to threaten the sustainability and effectiveness of SHGs; therefore, the future strategies and policies to implement those strategies for sustainable and effective functioning of SHG hold paramount importance.

7.3 Implications

Present study has revealed the existing scenario of self help group in Chhattisgarh. It looked into the differential effectiveness and functioning of SHGs formed under different programmes NABARD's SBLP, NRLM, IWMP, and ATMA, which would provide impetus to the future planning on empowerment of rural poor through SHG approach, particularly in the state of Chhattisgarh.

The present study revealed the characterization of different SHGs with reference to characteristics of group members, group features and the situational factors which influenced the group dynamics.

The differential pattern of group dynamics assessed through ten indicators provided an effective methodology to analyze the dynamism of SHGs on the basis of perceptions of the members on different indicators. This probing would be of great help to correct, if require, on particular issue(s) for betterment of group dynamics of respective SHGs.

The impact of SHG on livelihood security and level of livelihood have been viewed as most demanding and motivating factor to spread SHG approach; therefore, present study reflected the achievements made in this regard by different SHGs formed under different programmes.

Empowerment of rural poor in general and rural women in particular has been the key outcome of SHG movement in India since mid 1990s. The present study surfaced the empowerment of SHG members in selected region of

Chhattisgarh especially with reference to gender parity in production, resources, time, leadership and income domains.

The study provided a full insight into the constraints faced by the self help groups. Accordingly, a strategy is developed for mobilization and/or strengthening of effective and sustainable SHGs overcoming the existing constraints.

Moreover, this study would assume great significance in creating database for realistic planning and implementation of future SHG movement. Findings would also serve as a bench mark for future evaluation. A study like this would help in adding to existing store-house of knowledge concerning SHG approach and related issues. It would also guide future researchers in deriving insight in understanding many aspects relevant particularly to SHG approach adopted under different schemes implemented by various government and non-government organizations in general and Chhattisgarh State in particular.

7.4 Suggestions for Future Line of Work

The study was limited to one block of Kanker district of Chhattsigarh state including 12 SHGs. Hence, a detail study covering more SHGs, blocks and districts may be conducted in order to generalize the recommendations for the entire state of Chhattisgarh.

The activity-wise stratification of SHGs and study their differential performance may provide further impetus in research of SHG approach.

Self Help Groups are believed to bring about socio-economic empowerment of rural women by engaging them in productive economic activities leading to wealth creation, more expenditure, better monitory savings, *etc.* A study focusing on these aspects would help in verifying such believes.

Chapter 8

Sustainability of SHGs

In the study of rural development, there are three frameworks of analysis, viz. neo-classical, Marxian and systems approaches. While the latter two are holistic approaches, the neo-classical framework provides the underpinning of modern microfinance, with its focus on the decision making of individual economic agents towards (short-term) profit or utility maximisation. Neo-classical discussion and policy prescriptions in microfinance are directed towards creating the conditions for the operation of free rural financial markets. Institutional economists have further used the imperfect information paradigm to focus on transaction costs of credit delivery that serve to explain and justify high interest rates charged by village moneylenders (Tankha, 2002).

Within microfinance, sustainability can be viewed at several levels - institutional, group and individual - and can relate to organisational, managerial and financial aspects. However, it is the financial sustainability of microfinance institutions that has become the critical point of focus of mainstream analysis at the expense of the sustainability of the client/borrower. Gender empowerment represents another domain of influence of microfinance interventions. Mayoux (1998) provides three contrasting, but overlapping, paradigms of financial self-sustainability, poverty alleviation and feminist empowerment. Thus sustainability in the financial self-sustainability paradigm is seen in terms of financial self-sufficiency of the microfinance intermediary. Under the poverty alleviation paradigm sustainability is viewed in terms of long-term community self-reliance and self-determination of the poor, while the feminist empowerment paradigm aims at the development of self-sustaining women's organisations for long-term change in gender relations (Tankha, 2002).

The above discussion shows that the sustainability question has multiple dimensions that need to be reconciled in order to judge the appropriateness of different approaches. The sustainability question in respect of SHGs needs to be reviewed given the character, context and objectives of their promotion. The attractiveness of the SHG as a micro-bank serving its members arises from the low-cost retailing option it provides through externalisation of the transaction costs of financial intermediaries (in part through transfer of costs to the SHG and its members). However, as is increasingly becoming evident, only about one-third to one-half of SHG members are able to avail of loans out of external funds. Further, loan requirements and growth in demand for loans are constrained in many areas and even in the relatively better-off states where the bulk of the SHGs have been formed (Tankha, 2002).

It is instructive to note that the merits of the SHG as an on-lending group has its roots in the "flexibility" of the Rotating Savings and Credit Association (ROSCA) and the Accumulating Savings and Credit Association (ASCA). ROSCAs respond to changes in the tempo of trade and business activity and may die out and are resurrected, say, after a period of drought, war and other economic upheavals. Sustainability is not a necessary attribute of the ROSCA. Informal money management groups are short-lived and changing in their membership for good reasons - to maximise returns and safety. The desire of donors and MFIs to work with stable groups necessitates much investment in management style and leadership towards the development of sustainable structures based upon SHGs. In this context it is also relevant that the SHG bank-linkage programme was conceived as a supplementary programme to reach the poorest families not served by the banking system.

Finally, SHGs are being used in many states as building blocks for primary and secondary federations, financial and non-financial, to access loan funds and for the delivery of non- financial services. Where this has been done, the financial and organisational sustainability of the SHG federations comes into question as well (Tankha, 2002).

8.1 Sustainability of SHG Models

The issue of sustainability and the limited options available in different contexts has led to various "models" or strategies of SHG development that have been adopted by NGO promoters. The prospects for the sustainability of these strategies or paths, along with the available evidence, are discussed below.

The two broad paths for long-term SHG development in India were:

(i) SHGs directly linked to banks on a permanent basis

(ii) SHGs/federations of SHGs linked to various types of MFIs.

8.1.1 SHGs Linked to Banks

An SHG of 15 members, with a saving contribution of Rs. 30 per member per month can after one year save only Rs. 5,400 and be able to raise an initial bank loan of a like amount. However, with the accumulation of savings and the progressively increased leveraging of bank funds, this amount can increase rapidly. Even at this modest rate of saving and through internal rotation of funds, the SHG's savings fund can build up to about Rs. 30,000 after five years and it could be eligible for a loan of eight times this amount, or Rs. 2,40,000.

In practice only very few SHGs, apart from a few promoted by the best practice NGOs, have been able both to maintain their savings and to absorb loans on anywhere near this scale. Discussions with NGOs such as MYRADA, PRADAN, HCSSC, OUTREACH and ASSEFA, which have had the longest history of promoting SHGs, suggest that even in their programmes not more than one-third of their SHGs have been linked to banks and that only one-third to one-half of the members have borrowed from the SHG (Tankha, 2002). Over the comparatively long period of SHG savings promoted by these NGOs, average SHG savings range between Rs. 5,288 for HCSSC to Rs. 24,404 for MYRADA.

On the other hand, there is evidence that, apart from other benefits to members such as the rotation of savings for small-scale and emergency borrowings, SHG bank linkage reduces the transaction cost of both banks and borrowers for loans from the formal sector. There is no disputing the benefits of SHG-bank linkage in enabling access to loan funds of group members who were not being served by the banking system. However, the limits imposed the group's savings on SHG borrowings means that, for comparatively new SHGs only some of the SHG members can expect to receive a modest loan in the range of Rs. 3000 to 5000. Thus the majority of group members are either net savers or pure savers. Despite the critical importance of savings in the SHG model, it is the savings service that is least developed. There are invariably no savings products other than the compulsory weekly/bi- weekly/monthly contribution and open access to savings is not available. Profits of SHG operations are only shared in some organisations (Tankha, 2002).

Financial and organisational sustainability:

Given their good repayment performance, the viability or sustainability of SHGs in financial terms is currently not an issue. SHG income through interest charges and fines, particularly for absence and late attendance of meetings, though small is matched by an extremely low-cost of operations limited to maintenance of books of accounts and payment of an honorarium to the local accountant. Typically, borrowings are a 12% per annum under the bank linkage scheme and on lending to members at 2% per month flat rate. Indeed, well-functioning SHGs are able to use part of their profits for buying services of accountants, teachers, *etc.* from their own funds for other social and economic services.

The organisational sustainability of SHGs is more open to question. Little research has been done on the internal dynamics of SHGs, the access of relatively poor members to loans and the relative stake of, and costs borne and benefits realised by, different categories of members, e.g., the net borrowers, net savers and pure savers identified earlier. Experience suggests that even after a period of 3 to 5 years (the time usually taken for SHGs to achieve the experience and maturity required to function as an independent financial entity), SHGs are not equipped to engage directly with banks and other agencies (Tankha, 2002).

A senior manager of PRADAN reports a vulnerable stage of "group fatigue", two years or so after the formation of an SHG, when the initial enthusiasm of group functioning wears off and a period of renewed motivation is necessary. The building of the group cooperation and solidarity in non-financial activities, in opposition to the routine banking activity, can be seen as a must to reinforce the group cohesion. Thus while savings and credit may be used as an entry point activity, it may not be sufficient to ensure sustainable long- term functioning of the SHG.

Even where SHGs are not constituted into higher-level financial organisations, promoting NGOs have felt the need to form cluster level associations, for cross learning and for NGO monitoring during later stages of SHG development. HCSSC, PRADAN and MYRADA have encouraged SHGs to form into cluster bodies of 10-20 groups belonging to contiguous villages. In the case of other agencies which have helped to form member or community based organisations for financial services at the development block level, intermediate non- financial organizations at village or cluster level have been promoted with similar objectives. A related learning is that scattered SHG development is likely to be counterproductive and short-lived since each SHG needs to receive sustenance from wider associates even when not brought together into a higher order financial intermediary. This is especially so when the close contact and support of the promoting NGO is to be withdrawn (Tankha, 2002).

Quality of groups

The sustainability of SHGs is clearly related to the "quality" of groups promoted. A common characteristic of leading SHG promoters is the intensive training and capacity building undertaken at group level at various stages, which in turn contributes to higher costs of promotion. By now NGOs and banks have generally devised assessment criteria for appraisal and periodic evaluations of group performance and sustainability. Assessment indicators include frequency and attendance of meetings, volume of savings, rotation of own savings, development of financial and skills, quality of leadership, *etc*. It is common to find, though only at field office level, data on ratings of groups into good, average and poor or similar categories. Broadly speaking, even best practice NGOs generally have only about 50% of groups placed in the highest category, with 30-40% of SHGs needing additional support and 10-20% failing to take off. This accounts for the varying duration of external inputs necessary in SHG promotion and development and makes the case for continued long-term NGO presence in the area. The quality of SHGs promoted has become a serious issue in the case of SHGs promoted under the DWCRA, and other state initiatives, especially in Andhra Pradesh where limited and improper facilitation has led to a large proportion of SHGs becoming defunct. The revival of these SHGs is part of the activities of the CASHE project, APMAS and the ASP (Tankha, 2002).

Long-term prospects

The longer-term prospect for SHGs linked to banks is unclear. The leading NGOs covered in the study have phased out from some areas after having linked the SHGs formed to banks. There is, however, unease about their ability to continue to access to funds from the banking system and to move along a growth path out of poverty. With NGO withdrawal the monitoring of the SHGs also ceases and there is little information on their activities. However, since the NGO generally continues to work in the area it is in a position to undertake troubleshooting on their behalf.

The logical path for members of SHGs linked to banks should be graduate to (larger) individual loans under the bank's normal lending programme. This does not appear to be emerging, both on account of the absence of a vision at NGO and bank level as well as infrastructural and other constraints operative on the absorption of credit by poorer households (Tankha, 2002).

8.1.2 SHGs/Federations of SHGs Linked to MFIs

The sustainability of the SHG-based federations that have emerged for the four forms of links with financing institutions is considered.

(a) SHGs/SHG clusters/secondary federations linked to NGO-MFI

NGO lending to SHGs/individuals

In such a model with NGO staff effectively managing the loan portfolio, the role of the SHG, if it exists, is reduced to a facilitator, like the Grameen center composed of five or six joint liability groups. The model is suited to the agricultural labour class, which see less merit in coming together into SHGs, the preferred form of business and self-cultivating households (Tankha, 2002).

The advantage of this model is to largely de-link credit from savings as far as the SHG/individual are concerned and make possible larger loans to SHGs. On the other hand, NGO capacity for microfinance has to be built up and the NGO has to operate on a very small margin, of about 5-6% unless grant or subsidised funds are available. This necessitates large outreach. Though many small NGOs with less than 100 SHGs are keen to adopt this model, apart from other capacity building requirements, a minimum of 3,000 to 4,000 clients may be necessary for viable operations.

NGOs lending to SHG clusters/federations

Two concerns about this model remain. First, with the additional layer of the Cluster Level association (CLA), the cost of external funds to the SHGs increases. As against this SHGs could borrow under bank linkage at 12%. Second, the shift from the SHG to cluster as the unit of interface with bank/ NGO means that any organizational and financial weaknesses of the base SHGs are further magnified. For cluster associations to work successfully as independent MFIs they have to be based on viable SHGs with sound systems and financial skills developed through strong facilitation and capacity building by the NGO. Also, given the low outreach of CLAs, they are not in a position to realize economies of scale.

The other common and critical feature is the broad-based role for the federations that encompasses a range of largely economic functions centered on the development of livelihoods. Any assessment of the cluster as intermediary must take into account the additional activities and products introduced at cluster level that would not be possible while working with individual SHGs (Tankha, 2002).

(b) SHGs/SHG clusters/federations linked to not-for-profit companies and NBFCs

Under this model, however, SHG profits are not distributed to the members. While the SHG fund can be used to contribute to share capital of the company for increased availability of financial services, it is at the expense of returns on the contributions of individual members. It is argued, however, that this arrangement leads to lower rates on loans to individual borrowers since in the absence of distribution of profits SHGs do not attempt to increase their income through setting unreasonably high rates on loans to members (Tankha, 2002).

(c) SHG federations linked to wholesalers

This model is of multi-tier financial federations that access loans from wholesalers. Financial performance data of the multi-tier federation is not available. However, the more important question relates to the establishing and ensuring institutional sustainability of such federations capable of borrowing from apex bodies. Clearly, capacity building requirements are substantial. Thus while it may be possible for the community MF institutions to be built they need to rely on external, paid or unpaid professional services (Tankha, 2002).

(d) Mutually Aided Cooperative Societies (MACs)

The SHG model has been described as a village-banking model. Yet the SHG is not critical to formation of cooperative member-based organisations. While there has been a rush to register SHG associations as Mutually Aided Cooperative Thrift Society (MACTs) and NGOs and MF wholesalers are prepared to provide funds to MACTS for on lending to their members, doubts persist whether this is the appropriate model to build upon the SHG (Tankha, 2002).

First, given the member-owned structure of the MACTS, SHGs cannot strictly speaking be its constituents. However, through amendment of the by-laws of the MACTS individual members can have a share in the MACTS while the representative general body is composed of SHG leaders. In this institution the role of the SHG as a micro-bank is supplanted by the MACTS. The SHG becomes a facilitating institution rather than a fund manager – much like the Grameen centre – leading to its irrelevance or "disempowerment". Second, the MACTS does not overcome the SHG weakness of low degree of capitalisation towards mobilisation of funds. Third, MACTS generally continue to be managed and controlled by the NGOs supporting them. The chief functionary is almost invariably an NGO staff member. Finally, like all cooperative structures, the possibility of the exercise of political influence towards control of the MACTS cannot be ruled out. NGO support for MACTS follows a similar pattern to that of other federation types, with a planned tapering off of NGO grants and management inputs towards eventual self- sufficiency. However, there are no major instances of NGOs having phased out (Tankha, 2002).

Reports of the financial performance of MACTS present a mixed picture. In some accounts the savings constraint is highlighted – the absence of a strong savings mobilization effort through the SHG as a basis to source external borrowings. This has to be set against reports of an excessive accumulation of savings and poor off-take of loans in the absence of the available infrastructure support for SHG/MACTS members to undertake larger investments.

Nevertheless, the MACTS, whether formed at a small (cluster) level or as an apex of several clusters federations of SHGs at the mandal level, is being supported by agencies such as UNDP, CARE-India and the state government. NGOs are optimistic about the possibility of MACTS attaining sustainability after about three years or so of their support. At the same time the large-scale promotion of SHGs by state agencies have resulted in a situation where only about 20% of SHGs are functioning properly. The strengthening of SHGs, their federations and the formation of MACTS is a major exercise that is under way through the agency of the government-supported APMAS and with the support of DFID, CARE-India, BASIX and other major SHG promoters. One of the critical gaps that remain to be filled is that of a supportive infrastructure, particularly in the form of business development services and marketing linkages to enable the utilisation of the savings and the loan funds of SHGs and MACTS into productive investment (Tankha, 2002).

The wide spectrum of emerging models of SHG-based institutions reflects the efforts of NGOs and other stakeholders in addressing the challenge of sustainability. The multiple constraints and opportunities that determine the types of institutions promoted include:

(i) The legal and regulatory provisions in the states of operation;

(ii) The origins of the programmes and the broader vision for the community;

(iii) The capacity of the NGO to support microfinance;

(iv) Poverty contexts and social conditions in different areas;

(v) The availability of the physical and financial infrastructure and external support.

As with virtually all other institutional forms, the financial viability of SHG federations has not been conclusively demonstrated. However, as community based organisations with lower overheads and decentralised functioning they have the potential to provide cost-effective financial services once necessary investments are made to strengthen their management.

8.2 Rules and Regulations for Sustainable SHG

This is a collection from many SHGs and is used as a guideline for making rules and regulation for self help groups to follow.

Aims of SHG

To develop all round sustainable development of the members families, village and environment.

Objectives

1. To create appropriate awareness among the members for their all round development in the society.
2. To promote co-operation and self help attitude good habit among the members and voluntary collective working.
3. To promote savings attitude & habit among the members for their good future and to encourage the members to commit themselves to a regular savings system.
4. To meet the credit needs of the members in time for consumption, income generation, asset creation and other appropriate purposes.
5. To help members in acquiring and use of appropriate technical knowledge and managerial skills in relation to their occupations in order to increase the productivity.
6. To provide appropriate skill training for the members on need base and to help members for subsidiary income generation occupations.
7. To raise funds for the Group from appropriate sources and through group income generation activities.
8. To mobilize financial assistance, programmes and other resources from Banks, Govt., Voluntary Organizations and other sources and to take up various activities for the welfare of member families and the village.
9. To create awareness and work for human health development.
10. To improve the basic facilities for the village viz. drinking water, sanitation *etc*.
11. To promote sericulture forestry and encourage other environment improvement activities.
12. To encourage and help families for possession of alternate energy sources such as bio-gas, solar system *etc*., and also to help families to construct fuel efficient smokeless.
13. To run night class for literacy, numeracy and to improve the knowledge for all round development of people.
14. To mobilize and make available Animal Husbandry and Veterinary services in the village.
15. To give special emphasis for women and child development.

16. To arrange for the availability of improved agricultural equipments, implements *etc*. for the villagers and to run a service centre for the benefit of the members.
17. To sort out the disputes among the members or villagers.
18. To promote and encourage cultural, sports and other appropriate activities.

Rules and Regulations

(a) Members Admission

1. Only one responsible person from a poor family can become member in the group by paying the membership fee fixed by the Group in consensus.
2. The members have to pay the membership fee at the time of enrolment to the Group and the amount is not returnable under any circumstances as well as it will not be transferable to any of the legal heirs.
3. Persons who are involved in any party politics or involved in any type of anti-social activities or the wilful defaulters to any credit institutions or total dependents are not eligible for membership in the group.
4. The size of the Group shall be around 15 to 20 members.

(b) Group Meeting

1. The group meetings should be held once a week regularly on a convenient day, place and time as decided by the group.
2. The members should attend all the meetings in time without fail.
3. The member who unable to attend the meeting for genuine reason, the same has to be intimated to the Group in person or through a messenger in advance or at least informed at the same meeting. If fails the member is liable to pay fine for the absence as decided by the group.
4. The latecomers for the meeting and those who walk out in the middle without intimating the chairperson are liable to pay fine as decided by the group.
5. If a member was absent for three consecutive meetings without genuine reasons such person's membership shall be suspended or cancelled with or without notice.
6. Unrelated issues/points should not be brought for discussion in the Group meetings.

7. The members should not use vulgar words or physical force against any member during the meeting and the violators or misbehaviours have to pay fine as decided by the Group or their membership shall be cancelled.
8. The members should not stay separate from the Group during the meeting.
9. All members should have to sign in the minute's book at the end of the meeting after the recorded proceedings of each meeting are read out and confirmed.

(c) Members Participation in the Group

1. All the members should participate in the discussions and decision making process orally and mentally in the Group meetings.
2. Equal opportunity and encouragement should be given to all the members for their full participation in the meeting and in all the activities of the Group.
3. All the members should attend the related trainings/workshops/ seminars/ exposures *etc.* within and outside the village without fail. The violators are liable to pay fine except for the genuine reasons.
4. The members should co-operate and participate in all the developmental activities related to the member's families, village, the Group environment, *etc.* Appropriate actions against non-co-operators/non participators shall be taken by the Group.
5. The members should participate in researching/learning dissemination and adoption of appropriate technologies for development.
6. The illiterate members should show interest and put efforts to become literates.
7. The Group shall run a learning centre at the convenient time for this purpose.
8. All the members should involve in regular savings and credit management activities with a special focus.
9. All the members should work with concern towards creating/building socio-economic safety society and stress on population control.

(d) Duties and Responsibilities of the Members

1. All the members should promote and protect the co -operation and unity in the Group.
2. The members should create equal opportunities and give encouragement to all the members in the Group.
3. The members should mobilize, use and manage the needy resources properly/judicially.
4. It is the responsibility of all the members to take necessary collective action against the wilful defaulters and recover the loan amount.
5. The members should take responsibility carefully for their all-round development and should also take leadership responsibility in the Group with service motive.
6. All the members should involve in planning, implementing, monitoring and evaluate the development programmes of the family, village and environment time to time and to give attention for the results of evaluations with proper actions.
7. The members should promote and protect the unity and integrity of the group and the village.

(e) Executive Committee

1. An executive Committee consisting of three representatives selected unanimously in the Group should take overall responsibility of the smooth functioning of the Group.
2. The period of the executive committee shall be six months or one year as Group decides and the new committee should be selected three months in advance and trained to take over the position of their responsibilities.
3. The above three positions of the representatives shall be called as

   (i)...................................

   (ii)..................................

   (iii).................................

   Two out of the above three shall jointly operate the Group's Bank Account. This committee is responsible for the Group's cash at hand, cash at Bank or Post office.

4. The executive committee members should ensure remitting the cash at hand immediately to the Group's Bank Account. The cash at hand should not be kept more than two days, if kept, the concern member should pay fine plus interest as decides by the Group.
5. The executive committee members should ensure, proper maintenance of books of Accounts of the Group on day-to-day basis regularly up to date. They are also responsible for getting the Accounts inspected once a month and audited once in six months and to submit the accounts Statements and reports to the group meeting for appraisal.
6. The executive committee should facilitate the programme planning, timely implementation, monitoring, evaluation and actions.
7. The executive committee should facilitate the regular Group meetings and smooth functioning of the Group.
8. The executive committee members should have good contacts with the Govt. Departments, credit institutions, voluntary organizations and other related institutions and to mobilize resources for the improvement of Group and village.
9. The executive committee members can execute any agreement /deeds/ contract on behalf of the Group with prior discussions and resolutions in the Group each time for each subject.

(f) Members Savings in the Group

1. Each and every member should save at least defined amount per week or defined amount per month in the Group which should be maintained in each individual members name.
2. Savings amount may be withdrawn by the members only at critical circumstances with prior approval of the Group. However the members should maintain a minimum balance of Rs. 1000/- in their savings account.
3. No interest will be paid for the members savings with the group. But 12% interest shall be paid for the amount kept in the Group as fixed deposit for a minimum period of six months.
4. The members should save from the family income earned only, but not the borrowed money for interest.
5. The members will not be encouraged to adjust their savings amount against their loan due to the Group. Only at extreme circumstances the Group shall consider to adjust.

(g) Credit Management

1. The credit can be given to the needy members of the Group for the purposes such as consumption, income generation, asset creation, clearing the old burden loans, socio -religious and any other appropriate purposes.

2. The credit shall be given to the needy members of the group only after a careful study, through discussion and unanimous decision on the quantum of credit, rate of interest and the repayment schedule for each loan.

1. 3.The loaner should give a written agreement to the Group for the loan account as per the official procedures in the presence of the witness to the Group at the time of taking credit. The loaner should give a guarantor within the group if necessary.

4. The loan amount should be utilized for the agreed purpose only. In case of any change of purpose it should be approved by the Group in advance. The violators are liable to pay fine/penal rate of service charge.

5. The loaner should repay the credit/loan amount with service charge as per the repayment schedule agreed upon, if failed such member is liable to pay fine or penal rate of service charge @ Rs. 12% annum in addition to the normal rate of service charge from the date of overdue or as decides by the Group.

6. The members those who are irregular for the SHG meetings or irregular in savings are not eligible for credit/loan from the Group.

7. All the members shall have equal opportunity for loan from the Group on eligibility/priority basis.

8. The funds of the Group should be resolved to the optimum extent for the benefit of all the members.

9. The Group can avail loan from the Bank and other credit institutions, NGO's, other Groups and any other available sources for its activities and the same has to repay in time as per the terms and conditions agreed upon.

10. As well the Group can receive donations, grants, subsidy, and charity from Government NGO's other Organizations, individuals and any other available proper sources for the betterment of the group members' family, village and environment.

(h) Others

1. The members of the Group should maintain good discipline.
2. The members should bring along with them the members passbook for all the meetings of the Group.
3. All the members should abide by the rules and regulations existing and that may be formed from time to time.
4. All the members should refrain from all the bad habits.

8.3 Check List to Assess the Performance of SHG

It is important to know whether the SHG has been functioning well. The check list given below will help to assess each SHG in a simple, but effective manner.

S.No.	Factors to be checked	Very good	Good	Unsatisfactory
1.	Group Size	15 to 20	10 to 15	less than 10
2.	Type of members	Only very poor members	2 or 3 not very poor members	many not poor members
3.	Number of meetings	Four meetings in a month	Two meetings in a month	Less than two meetings in a month
4.	Timings of meetings	Night or after 6 p.m.	Morning between 7 and 9 a.m.	Other timings
5.	Attendance of members	More than 90%	70 to 90%	Less than 70%
6.	Participation of members	Very high level of participation	Medium level of participation	Low level of participation
7.	Savings collection within the group	Four times a month	Three times a month	Less than three times a month
8.	Amount to be saved	Fixed amount	Varying amounts	—
9.	Interest on internal loan	Depending upon the purpose	24 to 36%	More than 36%
10.	Utilisation of Savings amount by SHG	Fully used for loaning to members	Partly used for loaning	Poor utilisation
11.	Loan recoveries	More than 90%	70 to 90%	Less than 70%
12.	Maintenance of books	All books are regularly maintained and updated	Most important registers (minutes, savings, loans, *etc.*) are updated	Irregular in maintaining and updating books
13.	Accumulated savings	More than Rs. 5000/-	Rs. 3000-5000/-	Less than Rs. 3000/-

Contd.

14.	Knowledge of the rules of the SHG	Known to all	–	Not known to all
15.	Education level	More than 20 percent of members can read and write	–	Less than 20 per cent know to read and write
16.	Knowledge of Govt, programs	All are aware of Govt, programs	Most of the members know about Govt, programs	No one knows

1. SHGs with 12 to 16 "very good" factors can get loans immediately.
2. SHGs with 10 to 12 "very good" factors —need 3 to 6 months' time to improve, before loan is given.
3. SHGs with rating of less than 10 "very good" factors will not be considered for loan.

Source: A Handbook on Forming Self-Help Groups (SHGs), NABARD, Mumbai

Bibliography

Anjugam, M. and Ramasamy, C. (2007). Determinants of women's participation in self-help group (SHG) led microfinance programme in Tamil Nadu. *Agricultural Economics Research Review*, 20 (2): 283-298.

Anonymous. (2006). Valanmai Netkathir: women's self help groups- Avalur, Kancheepuram district. *Kisan World*, 33 (3): 11-12.

APMAS. (2006). Self help groups in India: a study of the lights and shades. EDA Rural Systems and Andhra Pradesh Mahila Abhivruddhi Society (APMAS). http://www.edarural.com/documents/SHG-Study/Executive-Summary.pdf.

APMAS and NABARD. (2009). Quality and sustainability of SHGs in Assam. Hyderabad: APMAS and NABARD. www.shgateway.in.

Arunkumar, T.D. (2004). Profile of SHGs and their contribution for livestock development in Karnataka. *M.Sc.(Agri.) Thesis*, University of Agricultural Sciences, Dharwad.

Asokan, R. (2005). Micro-enterprises: An alternative strategy for poverty alleviation. *Kisan World*, 32 (3): 49-50.

Balakrishnan, S. (2002). Micro-credit through co-operatives. *Kissan World*, 29(2): 58.

Banerjee, G.D. (2002). Evaluation study on self help group, financing agriculture - In house. *Journal of Agricultural Finance Corporation Ltd.*, 34 (2): 38.

Banerjee, S. and Dutta, A. (2014). Synergistic effects of microfinance through SHGs: A study of basic health and primary education indicators. *Microfinance, Risk-taking Behaviour and Rural Livelihood*, pp 113-129.

Banerjee, T. (2009). Economic impact of self-help groups -a case study. *Journal of Rural Development*, 28 (4): 451-467.

Barik, B.B. and Vannan. (2001). Promoting self help groups as sub-system of credit Cooperatives. *The Cooperator*, 38 (7): 305-311.

Barua, P.B. (2012). Impact of micro-finance on poverty: a study of twenty Self- Help Groups in Nalbari district, Assam. *Journal of Rural Development*, 31 (2): 223-244.

Bhagawati. (2006), Cooperative sector– changes required to operationally internalize concept of micro-finance. *The Cooperator*, 43 (8): 355-359.

Birner, R., and J. R. Anderson. (2007). How to make agricultural extension demand-driven? The case of India's agricultural policy. *IFPRI Discussion Paper 729*, Washington, D.C.: International Food Policy Research Institute.

Chambers, R. and Conway, G.R. (1992). Sustainable rural livelihoods: practical concepts for the 21st century. *IDS Discussion Paper 296*, Institute of Development Studies, Brighton, UK.

Coleman, B.E. (1999). The impact of group lending in northeast Thailand. *Journal of Development Economics*, 60: 105-141.

Dadhich, C.L. (2001). Microfinance – A panacea of poverty alleviation: A case study of Oriental Grameen Project in India. *Indian Journal of Agricultural Economics*, 56 (3): 419-424.

Darlingselvi, V. (2005). Impact of self help group training. *Kisan World*, 32(3): 31-32.

Das, S.K. (2012). Best practices of self help groups and women empowerment: a case of Barak Valley of Assam. *Far East Journal of Psychology and Business*, 7 (2): 29-51.

Dasaratharamaiah, K., Naidu, M.C. and Jayaraju, M. (2006). Women's empowerment through DWCRA – An empirical study. *Social Welfare*, 52 (12): 33-38.

Deininger, K. and Liu, Y. (2009). Longer-term economic impacts of self-help groups in India. *Policy Research Working Paper 4886*, The World Bank Development Research Group Sustainable Rural and Urban Development Team, Washington, D.C.: World Bank.

Devalatha, C.M. (2005). Profile study of women self help groups in Gadag district of Northern Karnataka. *M.Sc.(Agri.) Thesis*, University of Agricultural Sciences, Dharwad.

Department for International Development (DFID). (1999). Sustainable livelihoods guidance sheets, Eldis Document Store. http://www.eldis.org.

Dhara, A. and Mitra, N. (2005). Decentralised development and micro credit. *In: Decentralised Planning and Participating Rural Development,* Purnendu Sekhar Das (Ed.) New Delhi: Concept Publishing Company.

Dolli, S.S. (2006). Sustainability of natural resources management in watershed development project. *Ph.D.(Agri.) Thesis*, University of Agricultural Sciences, Dhaward.

Draper, N.R. and Smith, H. (1966). *Applied Regression Analysis*. John Wiley and Sons, New York.

Dwarakanath, H.D. (2001). Self-employment generation under DWCRA – A review. *Kurukshetra*, 49 (5): 33-41.

EDA Rural Systems and APMAS. (2006). The light and shades of SHGs in India, for CRS, USAID, CARE and GTZ/NABARD, CARE India.

Ganesamurthy, V., Radhakrishnan, M.K., Bhavaneswari, S. and Ganesan, A. (2000). A study of thrift and credit utilization pattern of SHG in Lakshmi Vilas Bank Suriyampalayam Branch, Erode. *Indian Journal of Marketing*, 12-16.

Ganesan, G. (2005). Rural transformation through self help groups (SHG). *Kisan World*, 32 (1): 13-14.

Ganesh, K.B. (2005). Brick maker becomes bricks kilns owner. *Kurukshetra*, 53 (10): 47-48.

Gangaiah, C., Nagaraja, B. and Vasudevulu Naidu, C. (2006). Impact of self help groups on income and employment: A case study. *Kurukshetra*, 54 (5): 18-23.

Garai, S., Mazumder, G., and Maiti, S. (2013). Group dynamics effectiveness among self-help groups in West Bengal. *Indian Research Journal of Extension Education*, 13 (1): 68-71.

Ghosh, S., Kumar, A., Nanda, P. and Anand, P.S.B. (2010). Group dynamics effectiveness of water user associations under different irrigation systems in an eastern India state. *Irrigation and Drainage*, 59: 559–574.

Gurumoorthy, T.R. (2000). Self-help groups: empower rural women. *Kurukshetra*, 48 (5): 36-39.

Harathi, R.A. (2005). Assessment of entrepreneurial activities promoted under NATP on empowerment of women in agriculture. *M.Sc. (Agri.) Thesis*, University of Agricultural Sciences, Dharwad.

Hashemi, S., Schuler, S., Riley, I. (1996). Rural credit programs and women's empowerment in Bangladesh. *World Development*, 24 (4): 635-653.

Hemalatha Prasad, C. (1995). Development of women and children in rural areas: Successful case studies. *Journal of Rural Development*, 14 (1): 65-87.

Hussain, B.Z., and Syed, Z.M. (2012). Impact of self-help groups on women empowerment in the union territory of Puducherry – a case study. *Marketing Zypher*, 1.

IFPRI. (2012). Women's empowerment in agriculture index. *Global Food Policy Report,* International Food Policy Research Institute (IFPRI), Washington, DC. http://www.ifpri.org/publication/womens-empowerment-agriculture-index.

Jain, R. (2003). Socio-economic impact through self help groups. *Yojana*, 47 (7): 11-12.

Jain, R. and Kushwaha, R.K. (2004). Self help groups and its impact on decision making. *Indian Research Journal of Extension Education*, 4 (3): 84-86.

Jha, P.N. (2004). Self help groups in accelerating transfer of technology for promoting sustainable agriculture and entrepreneurship development. *Indian Farming*, 54 (1): 36-38.

Joseph, L. and Easwaran, K. (2006). SHGs and tribal development in Mizoram. *Kurukshetra*, 54 (3): 37-48.

Joshi, M. (2006). Empowering Rural Women through water Shed Project. *In: Rural Women Empowerment,* Verma, S.B., Jiloka, S.K., and Khshwah, K.J. (Eds.). New Delhi: Deep & Deep Publications.

Kabeer, N. and Helzi N. (2005). *Social and Economic impacts of PRADAN's Self Help Group micro finance and livelihoods Promotion Programme: Analysis from Jharkhand, India.* http//www.lmp.act.org

Kala, G.S. (2004). Economic empowerment of women through self help groups. *Kisan World*, 31 (11): 25-26.

Kallur, M.S. (2001). Empowerment of women through NGOs: a case study of MYRADA Self-Help Groups of Chincholi Project, Gulbarga district, Karnataka State, *Indian Journal of Agriculture Economics*, 56 (3): 465.

Kamaraju, S. (2005). Self help groups – emerging rural enterprises. *Kisan World*, 32 (8): 25-26.

Kar, J. (2008). Improving economic position of women through microfinance: case of a backward area, Mayurbhanj-Orissa, India. *Indus Journal of Management & Social Sciences*. 2 (1): 15-28.

Kothal, K., Dongre, Y., Shridhara, T.N. and Moodithaya, M.S. (2003). Self help groups and the poor rural women. *Journal of Extension and Research*, 5 (2): 12- 13.

Krishnaiah, P. (2003). *Poverty Reduction: Self Help Group strategy*. New Delhi: UBSPD

Kumar, R. (2005). Sustaining rural development: micro finance in Haryana. *In: Strategies for Sustainable Rural Development,* Surat Singh (Ed.). New Delhi: Deep & Deep Publications. pp 268-281.

Kumar, S. (2010). Centre for Micro Finance Research & BIRD, Lucknow.

Kumaran, K.P. (1997). Self help groups: An alternative to institutional credit to poor: A case study of Andhra Pradesh. *Journal of Rural Development*, 16 (3): 515-530.

Lalitha, B.S. and Nagaraja, B.S. (2002). *Self help groups in rural development*, Dominant Publishers and Distributers, New Delhi.

Lalitha. B.S., and Nagarajan, B.S. (2004). Empowerment of rural women through Self Help Groups: a study in Tamil Nadu. *In: Empowerment of People.* R. Venkata Ravi, N. Narayana Reddy and M. Venkataramana (Eds.). New Delhi: Kaniska Publishers & Distributors. pp 73-85.

Lalneihzovi. (2007). *Women's development in India*, 1st edition, New Delhi: Mittal Publication.

Laxmi, R.K. (2000). Micro finance: new development paradigm for poor rural women. *Kurushetra*, 46 (14): 22-25.

Loganathan, P. (2004). SHGs and bank linkages. *Kisan World*, 31(4): 24-26.

Lokhande, M.A. (2013). Microfinance for women empowerment- a study of self help groups-bank linkage programme. *International Centre for Business Research Journal*, 2: 159-166.

Makandar, I.M. (2011). The role of self help groups and gender justice in India, In Lazer, Daniel, Aravanan and Deo, M. (Eds.). *Embodiment of Empowerment: Self Help Groups*, Vijay Nichole Imprints, Chennai.

Mandal, A. (2004). Swarnjayanti Gram Swarozgar Yojana and self help group: An assessment. *Kurukshetra*, 53 (3): 4-9.

Manimekalai, K. (2004). *Economic empowerment of women through self-help groups*. 1st edition, Concept, New Delhi.

Manimekalai, M. and Rajeswari, G (2001). Nature and performance of informal Self Help Groups: a case from Tamil Nadu, *Indian Journal of Agricultural Economics*, 56 (3): 453-454.

Meena, M.S., and Singh, K.M. (2012). Measurement of attitude and behavior of self-help group members: evaluative study of eastern India. Munich Personal RePEc Archive, Munich University Library, Germany.

Mohapatra, A. (2012). Empowerment of women at house-hold level through Self-Help Groups-a study of Khordha district of the state of Odisha, India. *International Journal of Research in Commerce, Economics & Managemen,.* 2 (5): 83-87.

Moyle, D. and Biswas. (2006). Personal and economic empowerment in rural Indian women: a Self-help Group approach. *International Journal of Rural Management,* 2 (2): 245-266.

NABARD. (2002). Ten Years of SHG-Bank Linkage: 1992-2002. NABARD and Micro Finance.

NABARD. (2005). Success stories from districts. *National Bank News Review*, 21(1): 24-26.

NABARD. (2011). Status of Microfinance 2009-10. Mumbai: NABARD.

NABARD. (2012). Status of microfinance in India 2011-12. Micro-credit Innovations Department, National Bank for Agriculture and Rural Development, India.

Namboodiri, N.V. and Shiyani, R.L. (2001). Potential role of self-help groups in rural financial deepening. *Indian Journal of Agricultural Economics*, 56 (3): 401-417.

Nanda, Y.C. (1999). Linking banks and self help groups in India and non-governmental organizations- Lesson learned and future prospects. *National Bank News Review*, 15 (3): 1-9.

Narang, U. (2012). Self help group: an effective approach to women empowerment in India. *International J. Social Science & Interdisciplinary Research*, 1 (8): 8-16.

Narasalagi, N.M. (1990). Study on profile of Mahila Mandals in Dharwad taluk of Dharwad district. *M.Sc.(Agri) Thesis*, University of Agricultural Sciences, Dharwad.

NCAER. (2008). Impact and Sustainability of SHG-Bank Linkage Programme. Submitted to GTZ–NABARD, Mumbai, India.

Nedumaran, S., Palanisami, K. and Swaminathan, L.P. (2001). Performance and impact of self-help groups in Tamil Nadu. Indian journal of Agricultural Economics, 56 (3): 471-472.

Nirmala, K.A. and Geetha, M. (2009). Socio-economic impact of microcredit: a study of measurement. *In: Micro credit and Rural Development,* Anil Kumar Thakur and Praveen Sharma (Eds.). New Delhi: Deep & Deep Publications. pp. 207-224.

Oommen, M.A. (2008). Micro finance and poverty alleviation: the case of Kerala's *Kudumbashree.* Working Paper No. 17. Centre for Centre for Socio-economic & Environmental Studies (CSES), Kerala

Pandian, P. and Eswaran, R. (2002). Empowerment of women through microcredit. *Yojana*, 46 (10): 14-16

Pankaj, N. (2001). Micro financing the self employment activities. *Kurukshetra*, 49 (10): 12-14

Paramasivan, C. (2013). Conceptual framework of women empowerment through SHG. *SELP Journal of Social Science*, 4 (17): 28-35.

Pfeiffer, J.W. and Jones, E.J. (1972). Annual Handbook of Group Facilities. Vol. 3, Pfeiffer & Company, San Diego, California, pp. 19-24.

Prasad, C.H. (1998). Implementation processes of women development programme (IFAD) – An experimental model. *Journal of Rural Development*, 17 (4): 779-791.

Pillai, T. and Nadarajan, S. (2010). Impact of microfinance - an empirical study on the attitude of SHG leaders in Kanyakumari district, Tamil Nadu. *International Journal of Enterprise and Innovation Management Studies,* 1 (3): 89-95.

Pitt, M.M, and Khandker, S.R. (1998). The impact of group-based credit programs on poor households in Bangladesh: Does the gender of participants matter? *Journal of Political Economics*, 106 (5): 958-96.

Prita, M.P. (2001). A study on the performance of self help groups in Dharwad district. *M.Sc. (Agri.) Thesis*, University of Agricultural Sciences, Dharwad.

Puhazhendhi, V. (2000). Evaluation study of self help group in Tamil Nadu. *National Bank for Agriculture and Rural Development*, Mumbai.

Puhazhendi, V. and Badatya, K.C. (2002). Self Help Group bank linkage programme for poor: an impact assessment. Paper presented at Seminar on Self Help Group - Bank Linkage programme at New Delhi.

Puhazhendhi, V. and Jayaraman, B. (1999). Increasing women's participation and employment generation among rural poor: An approach through informal groups. *National bank News Review*, 15: 55-62.

Puhazhendhi, V. and Satyasai, K.J.S. (2000). Economic and social empowerment of rural poor through self help groups. *Indian Journal of Agricultural Economics,* 56 (3): 450-451.

Puyalvannan. (2001). Micro finance and women empowerment through CRUSADE. *Kurushetra,* 49 (7): 41-45.

Raghavan, V.P. (2009). Micro credit and empowerment: a study of Kudumbashree projects in Kerala, India. *Journal of Rural Development*, 28 (4): 469-484.

Raheem, A. and Yasmeen, S.H. (2007). Empowerment of women through self help group: a view. *Kisan world*, 34 (3): 48-52.

Rajagopalan, M.R. (2007). Foreword. *In: Grass Root Entrepreneurship*, Lalitha N. (Ed.) New Delhi: Dominant Publishers and Distributors.

Rajendran, K. (2012). Micro finance through self help groups –a survey of recent literature in India. *International Journal of Marketing, Financial Services & Management Research*, 1 (12): 110-125.

Rajendran, K. and Raya. (2011). Does microfinance empower rural women?: a study in Vellore district, Tamil Nadu. *Indian Journal of Finance*, 5 (11): 47-55.

Rajivan, A. (2005). Micro credit and women's empowerment: a case study of SHARE micro finance limited. *In: Micro credit, poverty and Empowerment*, Neera Burra, Joy Deshmukh Ranadive and Ranjani K. Murthy (Eds.) New Delhi: Sage.

Ramachandran,T. and Balakrishnan, S. (2008). Self help groups on women's empowerment: a study in Kanyakumari district. *Kurushetra*, 57 (2): 31.

Ramakrishna, H., Khaja Mohinuddeen, J., Bibi Saleema, G.G. and Mallikarjuna, B. (2013). Performance of self help group-bank linkage programme (SBLP) in India – an analytical study. *Tactful Management Research Journal,* 1 (10): 1-6.

Rao, S.K. and Padmaja, G. (1998). Self Help Groups in Tirupati, Andhra Pradesh. *Social Welfare,* 45 (1): 25-27.

Rao, V.M. (2005). Building cooperative entrepreneurship through dairy cooperative in Ajmer. *The Cooperator*, 43 (5): 231-234.

Rao, Mohan, R.M., (2000). A study of women Self-help Groups in Andhra Pradesh. Thesis submitted to Andhra University, Visakhapatnam, and Retrieved from *ietd.inflibnet.ac.in/bitstream/10603/672/18/18_.pdf*, pp 243-253.

Rao,V.M. (2004). *Empowering Rural Women.* New Delhi: Anmol Publications.

Raveendran, N., Mathew, S., Ajjan, N. and Chinnaiyam, P. (2002). A comparative analysis of women's self help groups (SHGs) in Tamil Nadu and Kerala. *Indian Journal of Training and Development*, 32 (3): 133-138.

Reddy, C. S. (2005). SHGs: a keystone of micro finance in India: women empowerment & social security. http://www.aptsource.in/admin/... /1273818040_SHGs-keystone-paper.pdf .

Reji. EM. (2011). Micro finance and women empowerment: evidences from field study. *Journal of Rural development,* 30 (1): 25-44.

Roy, D. (2007). Mid-term Evaluation of the composition and working of Swarnajayanti Gram Swarozgar Yojana in 24 Parganas South District, West Bengal.

Sahu, G.B. (2010). SHG Bank Linkages in North West India: Experiences and Challenges in Financial Access and Poverty Alleviation. Centre for Micro-Finance, Institute of Development Studies, Jaipur, 5: 54-65.

Samar, K.D. and Raman, M. (2001). Can heterogeneity and social cohesion co-exist in self help groups? An evidence from group lending Andhra Pradesh in India. *Indian Journal of Agricultural Economics*, 56 (3): 389-400.

Sankaragoudar, S.K. (1991). Value orientation and socio economic characteristics of Lambani women of Dharwad district. *M.Sc. (Agri.) Thesis*, University of Agricultural Sciences, Dharwad.

Sarkar, S. and Baishya, S. (2012). Impact of microfinance in women empowerment of Assam, *BARNOLIPI - An Interdisciplinary Journal*, I (V): 46-69.

Sathiyabama and Saratha, M. (2011). Women empowerment and Self Help Groups in Mayiladuthurai block, Nagapattinam district, Tamil Nadu. *International Journal of Research in Commerce & Management*, 2 (9): 112-118.

Satish, P. (2001). Some issues in the formation of self help groups. *Indian Journal of Agricultural Economics*, 56 (3): 4710-417.

Satyasai, K.J.S. (2002). Micro finance in India: progress and perspective. *In: Institutional change in Indian Agriculture,* Suresh Pal, Mruthyunjaya, P.K. Joshi, Raka Saxena (Eds.) New Delhi: National Centre for Agricultural Economics and Policy Research. pp 305-320.

Savitha. (2004). Women empowerment on decision making in agriculture – An economic study in Mysore district. Karnataka, *M.Sc. (Agri.) Thesis*, University of Agricultural Sciences, Bangalore.

Selvachandra, M. (2004). Microfinance through self help. *Kisan World*, 31(12): 23-24.

Selvarajan, E. and Elango, R. (2004). *Rural Development: Programmes Externalities*. New Delhi: Serials Publications.

Senthil, V.K. and Sekar, V.S. (2004). Self help groups – A movement for women empowerment. *Kisan World*, 31(7): 13-14.

Sharada, O. (2001). Empowerment of rural women in SHG in prakasan district of Andhra Pradesh – An analysis. *M.Sc. (Agri.) Thesis*, University of Agricultural Sciences, Bangalore.

Sharma, P. and Verma, S.K. (2008). Women empowerment through entrepreneurial activities of self help groups. *Indian Research Journal of Extension Education*, 8 (1): 46-51.

Sharma, N., Wason, M., Singh, P., Padaria, R.N., Sangeetha, V. and Kumar, N., (2014). Effectiveness of SHGs in improving livelihood security and gender empowerment. Department of Agricultural Extension, Indian Agricultural Research Institute, New Delhi.

Shyledra, H.S. (2008). Role of self help groups. *Yojana*, January: 25.

Singh, K. (2001). Self help group – Its impact on the women folk of Manipur. *The Cooperator*, 39 (6): 315-317.

Singh., Y K., Kaushal, S. K. and Gautam, S.S. (2007). Performance of women's Self Help Groups in Moradabad district, U.P. *International Journal of Rural Studies*, 14 (2): 1-5.

Sinha, F. (2006). Self help groups in India: a study of the lights and shades. Hyderabad: A.P. Mahila Abhivruddhi Society

Srinivasan, G. and Rao, D.S.K. (1995). Financing of self help group by banks- some issues. *Working paper No.8,* Lucknow: BIRD

Subbiah, A. and Navaneeth, K.K. (2006). Linking self help groups (SHGs) with banks. *Kisan World,* 33 (5): 55-56.

Subramaniam, S. (2010). Empowerment of women through SHGs in Tirunelveli district, Tamil Nadu- a SWOT analysis, *Prabandhan: Indian Journal of Management*, 3 (3): 37-40.

Suguna, B. (2006). Empowerment of rural women through Self Hep Groups. New Delhi: Discovery Publishing House.

Surender, Kumari, S., Sehrawat, R.K. (2011). Can Self-Help Groups generate employment opportunity for rural poor? *European Journal of Social Sciences,* 19 (3): 371-379.

Suriakanthi, A. (2000). Literacy-essential for SHGs. *Social Welfare*, 47(6): 32-34.

Swain, R.B., and Fan Yang Wallentin (2007). Does micro finance empower women evidence from Self Help Groups in India. Uppsala University, Working paper 2007-24, August 2007. Retrieved from .http//www. nekuu.se./working papers.htm.

Tamizoli, P. (2004). Mainstreaming gender concerns in mangrove conservation and management: the Pichavaram coast, Tamil Nadu. *In: Livelihoods & Gender,* Sumi. Krishna. (Ed.) New Delhi: Sage publications. pp 92-108.

Tankha, A. (2002). Self-help Groups as Financial Intermediaries in India: Cost of Promotion, Sustainability and Impact. A study prepared for ICCO and Cordaid, The Netherlands. Sadhan, New Delhi.

Thejaswini, C.N., Chandrashekar, V. and Narayana Gowda, K. (2004). Performance of farm women in agriculture and income generating activities. *MANAGE Extension Research Review*, 5 (1): 68-73.

Thorat, Y.S.P. (2005). Microfinance in India: Sectoral issues and challenges. *National Bank News Review*, 21(1): 10-18.

Tripathy, K.K. (2004). Self-help groups: A catalyst of rural development. *Kurukshetra*, 53: 40-43.

Upadhye, V.V. and Kadam, N.L. (2008). A study of micro finance for women through SHGs in Miraj Tahsil in Sangli District in Maharashtra. In: Lazar, D. (Ed.). *Performance Evaluation and Enterprise Development*, pp. 428-429.

Usha Rani, Reddy, D.P.J. and Reddy, M.V.S. (2004). Women development: empowerment through Self Help Groups in Andhra Pradesh. *In: Communication and Empowerment of Women,* Kiran Prasad (Ed.) New Delhi: The Women Press. Vol.2, pp 616-624.

Vadivoo, Senthil, K. and Sekar, V. (2004). Self Help Groups- a movement for women services: how the poor India could be better served, *Kisan World,* 31 (7): 13-14.

Venkatachalam, A. and Jeyaprakash, A. (2004). Self-Help Group in Dindigul district, *Kisan World,* 31 (10): 29-30.

Vipinkumar, V.P. and Singh, B. (2001). A strategy for mobilization of an effective self-help group. *Indian Research Journal of Extension Education*, 1 (2): 20-26.

Vipinkumar, V.P. and Singh, B. (2002). Dimensions of self-help group dynamics of horticultural farmers. *Indian Research Journal of Extension Education*, 2 (1): 6-12.

Vipinkumar, V.P. and Asokan, P.K. (2014). A study of self-help group dynamics of women in Malabar fisheries sector. *Indian Research Journal of Extension Education*, 14 (2): 25-30.

VOICE. (2008). A report on the success and failure of SHG's in India – impediments and paradigm of success. Submitted to Planning Commission of India by Voluntary Operation in Community & Environment (VOICE), New Delhi.

Yunus, M. (2004). Grameen Bank, Micro Credit and Millennium Development Goals. *Economic and Political Weekly*, 39: 4077-4092.

NIPA Publications on Extension Education and Rural Development

S.No.	Title	Author	ISBN	Year
1	Advances and Challenges in Agricultural Extension and Rural Development	Rathakrishnan,T.	9789380235035	2009
2	Agricultural Extension: Worldwide Innovations	Saravanan, R.	9788189422967	2008
3	Agricultural Marketing: Perspectives and Potentials	Bhat, Anil	9789385516153	2016
4	Agrobiodiversity and Sustainable Rural Development	S.K. Soam	9789383305896	2015
5	Agro-enterprises for Rural Development and Livelihood Security	Sharma J.P. & S.K.Dubey	9789380235851	2011
6	Agroforestry Systems for Resource Conservation and Livelihood Security in Lower Himalays	Panwar,P & Dadhwal	9789381450215	2012
7	Agroforestry: Principles and Practices	Patra, Alok Kumar	9789381450765	2013
8	Agroforestry: Systems and Practices	Puri, Sunil & Pankaj Panwar	9788189422622	2007
9	Agroforestry: Systems and Prospects	C.B.Pandey & O.P.Chaturvedi	789381450970	2014
10	Ancestral Knowledge in Agri-Allied Science	Saha, Ratan Kumar	9789383305216	2014
11	Biotechnology in India: Initiatives and Accomplishments	Niladri Bag	9789385516252	2016
12	Computers in Agriculture: Fundamentals and Applications	Sharma Manish & Anil Bhatt	9789385516160	2015
13	Conducting An Effective and Successful Training Programme	Sontakki,Bharat S.	9789383305223	2015
14	Dimensions of Extension Education	Mohapatra, B.P.	9789381450987	2014
15	Disasters: Strengthening Community Mitigation and Preparedness	Khanna, B.K. & Neena Khanna	9789380235455	2011
16	Economics,Marketing and Sales of Agricultural Products	Thakur S. Nath	9789381450659	2013
17	Evaluation and Impact Assessment of Technologies and Developmental Activities in Agriculture,Fisheries and Allied Fields	Roy, A.K.	9789380235400	2011
18	Extension Education Management in Veterinary Sciences and Animal Husbandry	Singh, Ruchi	9789385516054	2016
19	Extension Management in the Information Age Initiatives and Impacts	Philip, H. & T.Rathakrishnan	9789381450543	2013
20	Extension of Technologies: From Labs to Farms	Anandaraja, N.	9788189422837	2008
21	Extension Techniques for Livestock Development	Singh S.K. & S.D.Singh	9789380235233	2011
22	Family Farming and Rural Economic Development	M.L.Choudhary & Aditya	9789383305858	2015
23	Farm Women Empowerment Through Dairy Co-operative Societies	Swain, Pitambar	9789383305186	2014
24	ICTs for Agricultural Extension: Global Experiments, Innovations and Experiences	Saravanan, R. ed.	9789380235240	2010
25	ICTs for Transfer of Technologies: Tools & Techniques	Verma, S.R.	9788193014479	2015
26	Information and Communication Technology for Agriculture and Rural Development	Saravanan, R.	9789380235882	2011
27	Information and Knowledge Management: Tools, Techniques and Practices	Roy, A.K.	9789381450628	2013

S.No.	Title	Author	ISBN	Year
28	Information Technology in Veterinary Science	Patil, Vivek	9788190851244	2009
29	Mobile Phones for Agricultural Extension	Saravanan,R.	9789383305230	2014
30	Question Bank in Extension Education	Mohammad,Asif	9788193014493	2015
31	Rural Livelihood and Food Security	Wani, M.H.	9789380235936	2012
32	Samekit Krishi Pranali	Kumar, Sanjeev	9789381450154	2012
33	SHGs in Techno-Economic Empowerment of Tribal Women	Satapathy, C. & Sabita Mishra	9789380235875	2011
34	Social-Ecological Diversity and Traditional Food Systems	Singh,Ranjay, Nancy J.Turner	9789383305360	2014
35	Technologies for Livelihood Enhancement	V.L. Chopra	9789383305810	2015
36	Traditional Agricultural Practices: Applications and Technical Implementations	Rathakrishnan,T.	9789380235028	2009
37	Women in Agriculture and Rural Development	Sridhara, Shakuntala: eds.	9788189422998	2009
38	Women in Sustainable Agriculture	C.Satapathy & Sabita Mishra	9789383305056	2014
39	Youth in Agriculture and Rural Development	K.Narayana Gowda:Ed.	9789381450758	2013

Zeitfracht Medien GmbH
Ferdinand-Jühlke-Straße 7
99095 Erfurt, Deutschland
produktsicherheit@kolibri360.de